T0134887

Intelligent Systems Reference Library

Volume 170

The aim of this series is to publish a Reference Library, including novel advances and developments in all aspects of Intelligent Systems in an easily accessible and well structured form. The series includes reference works, handbooks, compendia, textbooks, well-structured monographs, dictionaries, and encyclopedias. It contains well integrated knowledge and current information in the field of Intelligent Systems. The series covers the theory, applications, and design methods of Intelligent Systems. Virtually all disciplines such as engineering, computer science, avionics, business, e-commerce, environment, healthcare, physics and life science are included. The list of topics spans all the areas of modern intelligent systems such as: Ambient intelligence, Computational intelligence, Social intelligence, Computational neuroscience, Artificial life, Virtual society, Cognitive systems, DNA and immunity-based systems, e-Learning and teaching, Human-centred computing and Machine ethics, Intelligent control, Intelligent data analysis, Knowledge-based paradigms, Knowledge management, Intelligent agents, Intelligent decision making, Intelligent network security, Interactive entertainment, Learning paradigms, Recommender systems, Robotics and Mechatronics including human-machine teaming, Self-organizing and adaptive systems, Soft computing including Neural systems, Fuzzy systems, Evolutionary computing and the Fusion of these paradigms, Perception and Vision, Web intelligence and Multimedia.

** Indexing: The books of this series are submitted to ISI Web of Science, SCOPUS, DBLP and Springerlink.

More information about this series at http://www.springer.com/series/8578

Hariton Costin · Björn Schuller ·
Adina Magda Florea
Editors

Recent Advances in Intelligent Assistive Technologies: Paradigms and Applications

Editors
Hariton Costin
Faculty of Medical Bioengineering
University of Medicine and Pharmacy
Iași, Romania

Institute of Computer Science of Romanian
Academy, Iasi Branch
Iași, Romania

Björn Schuller
University of Augsburg
Augsburg, Germany

Adina Magda Florea
Faculty of Automatic Control
and Computers
University Politehnica of Bucharest
Bucharest, Romania

ISSN 1868-4394 ISSN 1868-4408 (electronic)
Intelligent Systems Reference Library
ISBN 978-3-030-30819-3 ISBN 978-3-030-30817-9 (eBook)
https://doi.org/10.1007/978-3-030-30817-9

This Springer imprint is published by the registered company Springer Nature Switzerland AG
The registered company address is: Gewerbestrasse 11, 6330 Cham, Switzerland

Preface

Intelligent assistive technology (IAT) is currently being developed, tested, and introduced worldwide, as an important tool to maintain independence and quality of life among community-living people with certain disabilities. This is very much in line with the European Union strategy for long-term care, which identified technologies as a key enabler for the aging population. IAT includes, e.g., sensor-based surveillance and monitoring systems, mobile technology such as wearable fall detectors, and activity bracelets as well as tablets with health information or alarm functions. Indeed, the application of IAT in care and services is a rapidly changing area, in which new products and services are constantly developed and introduced at a high pace. Ambient assisted living (AAL) technologies, among the most promising and fast-changing types of IAT, have been categorized into different "generations," according to how they have evolved over time. This categorization differentiates low-tech devices such as wearable alarms that only need user initiation (first generation) from systems for automatic detection of hazards (second generation) to more complex "smart" systems integrating home sensors and wearable devices (third generation).

Intelligent assistive technology bears huge potential in making a difference to the daily lives of millions of patients or individuals ideally in their everyday environments. Intelligent assistive technology encompasses a broad range of potential use cases and applications reaching from larger devices such as robots aiding patients in their movements or empowered by social skills for caretaking or even teaching individuals such as with autism spectrum condition to tiny portable wearables or even "invisibles" to be swallowed for 24/7 monitoring of our health and well-being. They mainly share two common factors: the intelligence side of such technology and the idea to improve the life quality of those concerned.

As to the intelligence side of matters, huge steps forward have been seen over the last half-decade or so, driven by factors such as the improvements reached by the advent and increasing advancement of "deep" learning and exploitation of "big" data. In addition, continuous improvements have also been seen in the development of sensing technology and energy awareness for the embedding of intelligent sensing (and actuating) on mobile platforms. The latter is of particular importance to increase the adherence of users, by reducing the efforts in "wiring up" and need to (repeatedly) charge devices, in the worst of cases even several times a day. Coming to energy awareness and embedding, "squeezing" of (large) and deep neural networks such as by intelligent and adaptive quantization of weights and activations or teacher networks that "teach" student networks with lower complexity have recently led to remarkable results, but in particular better exploitation of the novel improvements in mobile hardware tailored for such algorithms seems able to lead to the next generation of low energy consumption "green" mHealth and assistive technology devices.

Untouched by this, we are still far from using all information available when it comes to "mobile Health" (mHealth) sensing or to provide more intelligent assistive technology. In addition, as to the intelligence, many challenges are yet to be solved such as increasing the privacy of the data during transmission to servers. In fact, the ultimate goal is to best avoid such transmission by more efficient on-device processing, assuring that only a patient or user is the beholder of her/his data and its interpretations. Further, "white-boxing," i.e., increasing the explainability is an omnipresent requirement by users and providers alike these days, and only small steps such as the provision of meaningful confidence measures have so far been made. However, pointing for example back to the data points a decision made by an intelligent device comes from seems very needed. Further to that, drifting learning and classification targets with ever-changing manifestations of diseases and disorders have to be better handled including rapid adaptation and learning from a few examples. The latter certainly already benefits from the huge steps taken in transfer learning methods, which urges the community to provide more pretrained resources to transfer from. Such networks have first been seen at large in the computer vision community, and the computer audition and natural language processing ones. For other sensors or actuators and typical tasks in intelligent assistive technology, these are largely yet to follow. Furthermore, increasing usage of reinforcement learning to learn with the user in the loop or from thousands of users in parallel in the real-world application needs to be exploited more aggressively in real-life use cases given the huge potential it bears and has already proven in related cases such as automatic speech recognition. An alternative route has recently been offered by generative adversarial architectures, partially able to generate or "imagine" new training examples from the existent ones including a number of successes in this domain of interest. In particular typical to the domain of mHealth is also the huge challenge of coping with missing information due to sensor failure, or simply patients and users forgetting about their application. Recent machine learning offers its solutions such as latent convolutional neural networks, but further improvements will be required also on this end.

The main domains of health care in which IAT evolved during the time are:

- Prosthetics and Orthotics
- Assistive Devices for Persons with Severe Visual Impairments
- Assistive Devices for Persons with Severe Auditory Impairments
- Assistive Devices for Tactile Impairments

- Alternative and Augmentative Communication Devices
- Manipulation and Mobility Aids
- Recreational Assistive Devices
- Activities of daily living for older people.

For instance, in the last domain above one may say that one of the most desirable applications of IAT is as a tool to support older adults with dementia through activities of daily living (ADL), particularly because support for these tasks is often provided by informal, unpaid care. There are several studies where people with dementia reported the ability to independently complete ADL significantly affected his or her quality of life. Accordingly, the development of IAT for recognizing and supporting ADL has become a central research area with the goal of maintaining a person's ability to independently complete ADL as well as reducing the burden experienced by his or her caregiver.

User-centered design is a salient and desirable feature of the new IAT. Despite the wide range of AT in development to support people with certain disabilities or their caregivers, one may say that rather few of these IAT have really progressed beyond the initial development stage (e.g., a recent review of 58 technologies highlighted only eight devices that have undergone clinical trials and only two had entered real-world testing). As such, the appropriateness of many of these devices as tools to support the actual needs of people with disabilities and their caregivers remains in question. Recently, a small number of studies have looked specifically at (I)AT as tools to support ADL. For instance, a grounded qualitative analysis identified the category "daily activities" within the core theme "Problems in the Home" which found that the ADL most in need of support were dressing, taking medication, personal hygiene, food and drink tasks, and toileting. The varying results of these studies suggest that more information is required to generalize these needs into a foundation for the development of IAT for older adults with cognitive impairments.

In the book *Introduction to Biomedical Engineering*, edited by John Enderle, Joseph Bronzino, and Susan M. Blanchard, Prof. Andrew Szeto pointed out that in addition to avoiding common misconceptions, a rehabilitation engineer and technologist should follow several *principles* that have proven to be helpful in matching appropriate assistive technology to the person or consumer. Adherence to these principles will increase the likelihood that the resultant assistive technology will be welcomed and fully utilized.

Principle #1. The user's goals, needs, and tasks must be clearly defined, listed, and incorporated as early as possible in the intervention process. To avoid overlooking needs and goals, checklists and premade forms should be used.

Principle #2. The involvement of multidisciplinary rehabilitation teams, with differing skills and know-how, will maximize the probability of a successful outcome. Depending on the purpose and environment in which the assistive technology device will be used, a number of professionals should participate in the process of matching technology to a person's needs.

Principle #3. The user's preferences, cognitive and physical abilities and limitations, living situation, tolerance for technology, and probable changes in the future must be thoroughly assessed, analyzed, and quantified. Rehabilitation engineers will find that the highly descriptive vocabulary and qualitative language used by nontechnical professionals need to be translated into attributes that can be measured and quantified.

Principle #4. Careful and thorough consideration of available technology for meeting the user's needs must be carried out to avoid overlooking potentially useful solutions. Specific databases (e.g., assistive technology Web sites and Web sites of major technology vendors) can often provide the rehabilitation engineer or assistive technologist with an initial overview of potentially useful devices to prescribe, modify, and deliver to the consumer.

Principle #5. The user's preferences and choices must be considered in the selection of the assistive technology device. Surveys indicate that the main reason assistive technology is rejected or poorly utilized is inadequate consideration of the user's needs and preferences. Throughout the process of searching for appropriate technology, the ultimate consumer of that technology should be viewed as a partner and stakeholder rather than as a passive, disinterested recipient of services.

Principle #6. The assistive technology device must be customized and installed in the location and setting where it primarily will be used. Often seemingly minor or innocuous situations at the usage site can spell success or failure in the application of assistive technology.

Principle #7. Not only must the user be trained to use the assistive device, but also the attendants, caregivers, or family members must be made aware of the device's intended purpose, benefits, and limitations.

Principle #8. Follow-up, readjustment, and reassessment of the user's usage patterns and needs are necessary at periodic intervals. During the first 6 months following the delivery of the assistive technology device, the user and others in that environment learn to accommodate to the new device. As people and the environment change, what worked initially may become inappropriate, and the assistive device may need to be reconfigured or reoptimized. Periodic follow-up and adjustments will lessen technology abandonment and the resultant waste of time and resources.

However, once recent and further remaining obstacles are step-by-step overcome, we may soon witness a bright future of earlier than ever diagnoses available to most or all of us, hopefully seamlessly connected with intervention and rehabilitation programs perfectly tailored and attractive to the users for our all improved health and well-being.

Iași, Romania
Augsburg, Germany
Bucharest, Romania
June 2019

Hariton Costin
Björn Schuller
Adina Magda Florea

Contents

About the Editors

Prof. dr. eng. Hariton Costin B.Sc. in Electronics and Telecommunications (1980), Ph.D. in Applied Informatics (1998), MBA diploma, is a full professor at the University of Medicine and Pharmacy, Faculty of Medical Bioengineering, Iasi, Romania (www.umfiasi.ro). Also, he is a senior researcher at the Romanian Academy—Iasi Branch, Institute of Computer Science, within the Image Processing and Pattern Recognition Laboratory (http://iit.academiaromana-is.ro/personal/h_costin.html). Here, his studies are in image processing and analysis by using Artificial Intelligence methods, metaheuristic algorithms, and data fusion .

Competence areas include: medical electronics and instrumentation, biosignal and image processing and analysis, artificial intelligence (soft-computing, decision-aided systems), hybrid systems, HCI (human–computer interfaces), assistive technologies, telemedicine, and e-health. Scientific activity can be resumed by about 200 published papers in peer-reviewed journals and conference proceedings, eight books, four book chapters in foreign publishing houses, three patents, and two national awards.

Research activity: 33 annual research reports, a technical manager within FP5/INES 2001-32316 project, for a telemedicine application; responsible for the first Romanian pilot telemedical center in Iasi, director for eight national funded projects in bioengineering and (biomedical) image processing / analysis. He has served as the program committee member of various international conferences and reviewer for various international journals. Prof. Costin was invited as a postdoc

researcher at the University of Science and Technology of Lille (France, 2002, in medical imaging), at the University of Applied Sciences, Jena, Germany (2013, in biosignal processing) and had invited talks at international conferences. Prof. Costin is a senior member of the IEEE/Engineering in Medicine and Biology Society (EMBS) and of other five scientific societies. Also, he is general chair of the IEEE EHB conference series (2007–2019), i.e., E-Health and Bioengineering (www.ehbconference.ro).
e-mail: hcostin@gmail.com; hncostin@mail.umfiasi.ro

Björn Schuller received his diploma in 1999, his doctoral degree in 2006, and his habilitation and was entitled Adjunct Teaching Professor in Signal Processing and Machine Intelligence in 2012 all in electrical engineering and information technology from TUM in Munich, Germany. Since 2017, he is Full Professor and Chair of Embedded Intelligence for Health Care and Wellbeing at the University of Augsburg, Germany, in the Faculty of Applied Informatics and the Faculty of Medicine. At the same time, he is Professor of Artificial Intelligence in the Department of Computing at Imperial College London, UK, since 2018, previously being a Reader in Machine Learning since 2015 and Senior Lecturer since 2013. Further, he is the co-founding CEO and current CSO of audEERING GmbH. In 2019, he was appointed as Honourary Dean of the Centre for Affective Intelligence at Tianjin Normal University, Tianjin, P.R. China. In 2014–17, he was Full Professor and Chair of Complex and Intelligent Systems at the University of Passau, Germany, where he previously headed the Chair of Sensor Systems in 2013. In 2006–14, he headed the Machine Intelligence and Signal Processing Junior Group at TUM. In 2013, he was also invited as a permanent Visiting Professor at the Harbin Institute of Technology, Harbin, P.R. China, and a Visiting Professor at the Université de Genève in Geneva, Switzerland, in the Centre Interfacultaire en Sciences Affectives and remained an appointed associate of the institute, and since 2018, he holds also a position as Professor of the Vilnius Gediminas Technical University, Lithuania. In 2012, he was with

Joanneum Research, Institute for Information and Communication Technologies in Graz, Austria, remaining an expert consultant of the institute 2013–2017. From 2009 to 2010, he was with the CNRS LIMSI Spoken Language Processing Group in Orsay, France, and was a visiting scientist at Imperial College. Best known are his works advancing Machine and Deep Learning for Multimodal Human Data Analysis and its application in Health and Affective Computing. Dr Schuller is Fellow of the IEEE and Senior Member of the ACM. Before, he was President of the Association for the Advancement of Affective Computing (AAAC). He (co-)authored five books and more than 800 publications leading to >20,000 citations (h-index = 73). He was Editor in Chief of the IEEE Transactions on Affective Computing, General Chair of ACII 2019 and ACM ICMI 2014, and a program chair of INTERSPEECH 2019, ACM ICMI 2019 and 2013, IEEE SocialCom 2012, and ACII 2011 and 2015. In 2019, he received the IEEE Computer Society Golden Core Award and was listed in "100 Top AI Leaders in AI in Drug Discovery and Advanced Healthcare" by Deep Knowledge Analytics. In 2015, he has been honored as one of 40 extraordinary scientists under the age of 40 by the World Economic Forum. He is also a consultant of global enterprises such as BARCLAYS, GN Store Nord A/S, HUAWEI, and SAMSUNG.

Adina Magda Florea is professor of Artificial Intelligence at the Department of Computer Science of University POLITEHNICA of Bucharest and the Director of the Artificial Intelligence and Multi-Agent Systems Laboratory (https://urldefense.proofpoint.com/v2/url?u=https-3A__aimas.cs.pub.ro_&d=DwIFaQ&c=vh6FgFnduejNhPPD0fl_yRaSfZy8CWbWnIf4XJhSqx8&r=tXng7JtbiNZ_fPY83slTNeAj9_8jitOIwSlgY9ixQ3lGzu3-pKEwIHmSQpu2-iSg&m=rfeYEkzr9zEsyOCffB1mSVkVZiKOG-s6wi0blTbO8NQ&s=gAM5ws8hSewkfc9VAGumgd64fFwSIOjwb2uW9SsXNUE&e=). She is the Dean of the Faculty of Automatic Control and Computers of POLITEHNICA. Professor Florea is Senior Member of IEEE, Senior Member of ACM, and President of the Romanian Association for Artificial Intelligence. Her research interests are in multi-agent systems, machine learning, social robots and human-robot interaction, and ambient intelligence. She is the author of over 200 scientific papers in journals and conferences, was member of over 170 conferences programme committees, and director of over 20 national and international grants (National Research Programmes, FP6, FP7, AAL, COST, Minerva, Erasmus Mundus, Structural Funds).

Chapter 1
Cloud and Internet of Things Technologies for Supporting In-House Informal Caregivers: A Conceptual Architecture

Antonio Martinez-Millana, Gema Ibanez-Sanchez and Vicente Traver

Abstract Persons in a situation of dependency, or independent but with deficiencies in their autonomy, have specific needs for a better management of their long-term care. New sensing technologies based on real-time location systems, mobile apps, the Internet-of-Things (IoT) paradigm and cloud systems can be used to collect and process information about their activity and their environment in a continuous and truthful way. In this chapter, we analyse current solutions available to support informal caregivers and propose an innovative framework based on the integration of existing IoT products and services of cloud architectures. From the technological point of view, the system we propose is focused on the integration and combination of technologies for providing support for the informal caregiver in long-term care. The differential factor of these technologies is the customization level according to the specific context of the end-users. The main contribution of the proposed systems relies on the intelligence and the management of recorded events to create complex and reliable alerts, and its ability to configure multiple end-user instances and configurations (e.g.: needs, countries, regions, cultures). These type of systems should be sustainable and efficient, and that is why the inclusion of cloud technologies can grant its flexibility and scalability.

Keywords Dependency · Internet of Things · Cloud systems · Ambient assisted living · Informal caregiver · Apps · Web portal · Ageing

1.1 Introduction

Empowering and motivating people in the society are two major societal challenges which become crucial when it comes to people with special needs [3]. It is a well-known fact that our world is ageing rapidly [28] and living longer implies the risk of age related impairments that reduce significantly the Quality of Life (QoL). Elderly people should be supported on improving and maintaining their independence, func-

A. Martinez-Millana (✉) · G. Ibanez-Sanchez · V. Traver
ITACA, Universitat Politècnica de València, Camino de Vera sn, 46022 Valencia, Spain
e-mail: anmarmil@itaca.upv.es

H. Costin et al. (eds.), *Recent Advances in Intelligent Assistive Technologies: Paradigms and Applications*, Intelligent Systems Reference Library 170,
https://doi.org/10.1007/978-3-030-30817-9_1

tional capacity, health status as well as their physical, cognitive, mental and social wellbeing [10]. This challenge has been already approached by the European Union with actions like the European Innovation Partnership on Active and Healthy Ageing (EIP-AHA) and some generic frameworks have been created [2].

The current socio-economic model demands critical cutting-edge research for ensuring sustainability. Guidance and empowerment of population may facilitate facing the challenges of getting older in terms of health parameters [8, 15]. Among the several factors influencing ageing condition (e.g.: sleeping, physical activity, nutrition, workload…), there is clinical evidence that maintaining people's health conditions can only be ensured by keeping in perfect shape the triangle of Rest, Nutrition and Physical Activity [6, 23]. However, people need more than rules and knowledge to be healthy. Social and health care systems are tackling this change by promoting empowerment and motivation of ageing society. However, these two streams must be coupled with other services and interventions to bring about large changes in habitual lifestyle, like the support informal caregivers can bring [33].

Global population is getting older [29]. Between 2020 and 2030, the number of people aged 60+ years is expected to grow from 901 million to 1.4 billion. By 2050, it is expected to be doubled to nearly 2.1 billion. Projections indicate that in 2050 people aged 80+ will reach 434 million, which is three times the current size of people over 80 years old. Although most of older persons will live in good health, prevalence of disability will increase proportionally, being a huge burden for health care systems.

Changes are needed to adapt health and social systems to maximize health and well-being at all ages. The National Academy of Sciences states that the changes needed should not imply to increase substantially budgets but the democratization of technology-based systems to help in the creation of sustainable health care systems [35]. An increase in life expectancy age entails the apparition of new physiological changes which increase the risk of chronic disease. By age 60, the major burdens of disability and death arise from age related losses in seeing, hearing and moving, and non-communicable diseases, including heart disease, stroke, chronic respiratory disorders, cancer or dementia. Strategies to reduce the burden of disability and mortality in the elderly by enabling healthy behaviours can start early in life and should continue across the life course, helping in preventing these diseases. The gross charge of the European Long-Term Care (LTC) is allocated to informal caregivers, although up-to-date statistics in the European Union limited [20]. The reasons of this situation are the lack of commonly agreed definitions and the fact that informal care is mostly delivered at home through informal arrangements which are not even recognised as such (a spouse caring for her/his dependent partner) and are inherently difficult to measure. Besides, the involvement of (mainly female) migrants in this process, albeit significant in Europe and elsewhere and widely acknowledged, is even less explored and raises further measurement complexities from different dimensions.

In this context, Information and Communication technologies (ICTs) can contribute to support patients to manage their disease and to promote a healthy ageing, but also to support professional and informal caregivers for the LTC. The objective of this chapter is to introduce a system based on ICT and the Cloud technologies to support the person in a dependent situation in their autonomy and the work of informal caregivers. The system has been proposed on the basis of the current state of the art in the technologies for supporting informal caregivers and the cutting edge of applications in the new era of communication (cloud computing and the internet of things).

1.2 Background

From the technological point of view, we focus the innovation of the system in the development of a comprehensive solution, which combines a set of existing technologies not yet integrated in the field of support for the informal caregiver in LTC. The most powerful part from the point of view of pure research and development focuses on the ability of the system to trigger complex and reliable alerts, and its ability to configure multiple client instances in multiple configurations (e.g.: countries, regions and cultures).

The main drawbacks of such system is that commercial technologies are not ready to be deployed in the Cloud and must be adapted to benefit from their advantages [37]. These type of environments can help to first, gain efficiency and ease the deployment, and second, adapt the process of caring for people needing assistance and its management, including specific algorithms for this domain. In this section, we describe the state of the art of technologies to enhance the innovation of the proposed solution

1.2.1 Cloud Technologies

According to the report "Cloud Computing Challenges and Opportunities" of the National Observatory of Telecommunications and the Information Society (ONTSI) May 3, 2012, the adoption of cloud technologies by Small and Medium Enterprises (SMEs) has its limitations and strengths [11]. In particular, the report confirms that the cloud technology is a key element for the commercialization of services, internationalization, productive and financial efficiency, quality implementation and innovation. More in detail, and related to the scope of informal caregivers, Cloud paradigm has more benefits than limitations when compared to a classic hosting model:

- Economic and financial advantages: the fact of migrating systems to the cloud means an economic saving in licenses, in investment and in infrastructure since the Cloud may not have to be owned by the system holder (public model), and especially thanks to the pay-per-use model. Cloud involves the application of economies of scale models to the virtualization of computing infrastructures [41].

- Outsourcing of operations: the companies delegate the management of the infrastructure to the supplier, which allows the company to concentrate their efforts on its own business.
- Flexibility: hiring and the management of resources, as well as its total or partial deactivation, is practically immediate. Such systems should be scalable, and in the case of assistive technologies this is a key aspect, since it is expected a sustained growth in the number of clients and not only in a single geographical area. It is important to be able to configure the Cloud itself to be able to have more resources on demand at specific times of the day with more algorithmic load and less in others in order to optimize the resources contracted [39].
- Security and permanent updating: relying on external management ensures that the outsourced technologies are be continuously updated and managed so that the solutions are be up to date in security patches. This also allows a constant source of opportunities thanks to the early adoption of the solutions [16].
- Efficiency: from the point of view of a technological platform hosted in a server (e.g.: monolithic) to a cloud provider or cloud technology has the following advantages:
  - Ability to manage multiple instances efficiently from a single command panel thanks to the abstraction of the layers of the infrastructure. One instance of the system by region or country, or by regions with different personal data management needs.
  - Ability to properly manage the load of the applications that are executed in different nodes, which results in a good functioning of the system before peak loads.
  - Ability to update and evolve all instances of the system in a few steps.
- Access to multiple applications and solutions: for example, one of the SaaS services (Software as a Service), which would allow more costs to be saved in a context of multiple subscribers to one or several services, for example access to a CRM (Customer Relationship Management) software [32].

From the point of view of the end users that are going to make use of the system, the Cloud deployment entails the system to be a pervasive application at the service of citizens, as Gmail can be today [9]. In this case, the system can be the Cloud platform for informal caregiver support in the LTC of people assisted. These services are based on [21]:

- The availability of data in the cloud accessible at any time or place and from almost any terminal.
- Secure storage and management given the high sensitive level of the data to be stored.
- Intelligence of the system in the cloud, which allows not having to have complex systems in the home that have to be managed but that relies on remote analysis, which now costs are very reduced thanks to scale economies.

In a Cloud environment, the type of services offered are of utmost importance. The Cloud Computing model groups services into three layers (Fig. 1.1): Software

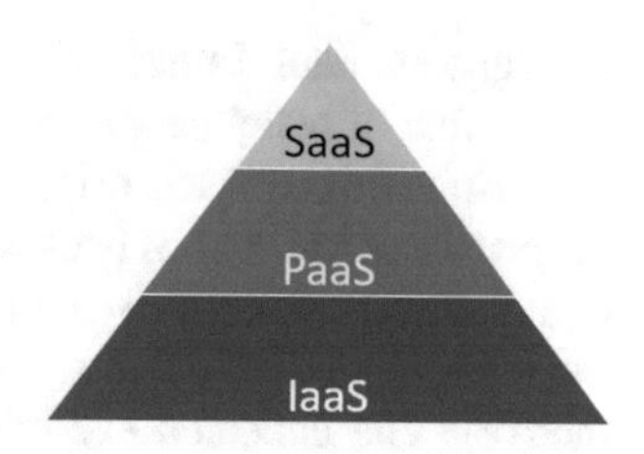

Fig. 1.1 Cloud computing pyramid

as a Service, Platform as a Service and Infrastructure as a Service [27]. More than layers, this classification finds different levels of abstraction of the cloud services, or simply different types of Cloud services and for different purposes depending on how deep the system wants to delve into the infrastructure and its virtualization:

- Software as a Service (SaaS): It is considered the top layer as it hides all the complexity of the cloud that is under it. This layer offers applications to be used directly by users either in a corporate environment or by clients of the applications that we bring to the cloud. It is the most visible layer for most of the citizens who use the cloud concept and Web 2.0 tools today. The software is instantiated in one place and not in the users' machines. Some examples are Google Docs, CRM, etc. [34].
- Platform as a Service (PaaS): This is a set of resources accessible by programmers as a programming and testing environment during the development phase and subsequent deployments. It means having some control by the client of what resources to install for the development and deployment of a system, without taking into account where or how these resources are supported. Some examples are Integrated Application Servers, Database management systems, template portals (without content), Application life cycle management, Internal messaging services, etc. [30].
- Infrastructure as a Service (IaaS): this is the lowest level of the Cloud model, directly accessible to the management of the concrete infrastructure and the abstraction that the cloud model introduces. Some examples of IaaS are computing, that is, the ability to control which processes are executed and with what availability of resources, Access to data storage to assess where the data is stored and in which format, especially if it is sensitive data (e.g.: health related data) [24].

Different cloud service types have been considered for the proposed architecture. The objective was to propose a solution, which can be integrated with existing and legacy services for informal caregivers.

1.2.2 Internet of Things

The concept of the Internet of Things has the same philosophical inception of the environmental intelligence, that is, the democratization of computing to stop being a unique element for each person to be an unlimited set of devices to serve not only of one fate of many people, as Mark Weiser predicted [14].

Despite this common origin, Internet of Things has a particular character, since the origin as a concept goes back to the time when RF-ID exploded as technology at the end of the 90s of the 20th century. The RF-ID must be understood as a technology for the identification of objects, either by excitation (passive) or by one's own initiative (active). The expected evolution of the RF-ID was to label any physical object, whether for professional applications such as logistics, to more popular applications such as games or food control [38].

Relying on the concept of identification, there subsequent step was to consider the communication of the objects between them. Furthermore, what if these objects are not only capable of issuing an identifier but also capable of providing functionality? What if they are communicating with each other, and moreover they do it globally through the Internet? By answering these questions is how the Strategic Research Agenda of the European Project Cluster on the Internet of Things [19] defines:

The Cluster for European Research Projects in the Internet of Things (CERP-IoT) defined an IoT networks as a "*Dynamic network infrastructure with self-configuration capability and based on standard and interoperable protocols in which physical and virtual things have identities, physical attributes, virtual personality and use intelligent interfaces, and are integrated in a transparent manner in the global information network.*"

Some examples of Internet of Things applications and services for ageing population are:

- Nabaztag by Violet. A family of personal devices integrated in a central unit with the appearance of a rabbit. The rabbit itself is the center of the applications that are deployed at the home of the user. It is connected to the internet, a power cable, 1 button, speaker and microphone, 3 lights that can change color independently or in a coordinated manner, and two motors that make their two ears move circularly. The user configures their capabilities by accessing a website. Two rabbits connected anywhere in the world can be paired so that the movement of one ear of one, causes an automatic interaction in the other, for example, the movement of the ear in the other. The rabbit is also able to recognize some objects tagged with RFID and give them "life": for example, when approaching a tale that bears an NFC tag, the rabbit is able to narrate it (http://www.nabaztaglives.com/).
- Xively. It is a website upload sensor data. The data is stored using a proprietary API of the provider. Once uploaded, we can consult them with different reasons. Although the normal thing is to use a Gateway (like a PC) that uploads the values to the cloud, there are hardware devices compatible with the API that upload the values directly to the website. From the site it is possible to make the data public, accessible by third parties, private, search, statistics, analyze trends … There are free modes of use and payment according to traffic. In this application the IoT aspect is combined with the Cloud (https://xively.com/).

– Smart-glasses. An IoT-based elderly behavioral difference warning system, which consists of wearable smart glasses, a Bluetooth Low Energy based indoor trilateration position, and a cloud-based platform [5].

1.2.3 Smart Rules Systems

The intelligence of a system can be described as a set of conditions, rules, policies and restrictions that share a set of internal parameters (known only by the intelligence system itself) and external parameters (whose generation and/or modification depends the measured variables, captured events captured or user input) [17].

In this context, a fact is a unit of information that describes a modelled concept of reality, defined by an identification and a series of valid attributes throughout the life of the same. Facts can be modified at any time in response to a change in their properties.

An event is a unit of information that presents the same characteristics of a fact and also includes a temporary reference. This reference is described by the time instant in which the event occurs and the period where it is valid (the time during the event it describes is true).

Finally, a rule is a minimum conditional structure that controls the generation, modification and elimination of events and/or events, as well as the control of inputs and outputs of the intelligence system based on compliance with a series of patterns normally defined by logical relations, mathematics or temporary events, events and constants present in the system.

Drools is a rule management system owned by JBoss Inc. that provides different solutions to problems related to the management, analysis, automation and optimization of business processes based on the use of a rule engine. Drools includes solutions for the complete management of the rules engine, both in the case of expert systems modelling (systems specialized in making decisions based on sufficient and precise knowledge of reality) or complex event processing systems (oriented to the management in real time of events and events). The advantages of using Drools with respect to other rule management systems are the following: (1) it is a project with a high frequency of updating and inclusion of improvements (more than two annual releases). (2) It is widely using and predominant in commercial applications. (3) Follows Java EE standards that ensure interoperability with most enterprise software solutions. (4) Good temporal response of the RETE00 decision making algorithm and stable and scalable memory consumption and (5) Complete and continuous improvement API that allows carrying out management operations through different methods, guaranteeing a good modularity and integration capacity.

1.3 Results

1.3.1 Informal Caregiver Technologies

Technologies for supporting informal caregiving are into a continuous development stage, especially in Europe [36]. However according to a recent review, there still exists a big gap existing in enabling older people and their informal caregivers to better understand smart home monitoring information [4]. Nowadays, almost all informal caregiver support initiatives focus on providing them with tools through training and access to knowledge in order to better face the problem for which they are not prepared. Table 1.1 shows the results of existing projects and initiatives aimed at the support of informal caregivers in long-term care in a European context. Almost all the actions that use information technologies are based on 2.0 websites that try to give information in a more or less organized way, about the care of the elderly.

The care of a person in a situation of dependency can be sometimes stressful and confusing; therefore, in the market have appeared an increasing number of applications for smartphones that pretend help informal caregivers to keep track of medication doses, nutritional requirements and other daily care needs of the people assisted.

Table 1.2 shows some of the solutions aimed at supporting the informal caregivers for Smartphone in the market.

1.3.2 Opportunities of Cloud Computing for Supporting Informal Caregivers

The system should include the appropriate tools to manage each of the instances of the global solution deployed and maintained by the system owner. This functional group is called the back-end. In the back-end it should be necessary to implement tools to guarantee dynamic scalability according to the load of calls to the system. It is expected that in an area of application to people assisted in which aspects of behaviour are monitored many individuals from the client population perform tasks at similar times of the day while in others they do not generate information. For example, when getting up all the assisted people, there will be rush time spans between 7.30 and 9 in the morning [18]. Another problem is to guarantee the serialization (the data will arrive sequentially for each and every one of the deployed houses) of the data receptions for a single assisted person, to maintain the consistency in the analysis algorithms.

Another challenge is the configuration of the data layer, which should ensure their redundancy and security, as well as being able to configure in an agile way the passage of data between different instances (allow the mobility of users between instances). Here the work would focus on assessing what kind of storage service and what kind of Service Level Agreement is established with the provider (for example, Azzure, Amazon, Gae). System owners should be able to efficiently manage the

Table 1.1 Technologies to support informal caregivers

Name	Type of technology	Target group	Endpoints	Status	Countries
SOPHIA	Telephone, sensors, video	Persons needing assistance	Communication	Pilot (finalized)	DE
Red Cross Spain	Panic button and network communication	Elderly and emergency call center	Emergency care	Deployed	Europe
ACTIVAGE	IoT Ecosystem	Professionals or ICs	Active and healthy ageing	Under ploting	SP, IT, GR, UK, DE, FI, FR
Pflegewiki Carers Direct Helpline *Ser cuidador*	Telephone and/or website	Informal caregivers	On-line information and LTC guidance	Deployed	DE, UK, SP
Carers UK15 SEKIS16 *Cuidadoras en red*	Web portal 2.0 (featuring forums, chats, blogs, etc.)	Informal caregivers	Peer support, mutual assistance and information exchange	Deployed	UK, DE, SP
IRDI	Blog	Informal caregivers	Social network, news	Deployed	SP
Sercuidador/a http://www.sercuidador.es/	Web portal	Informal caregivers	Social network, news	Deployed	SP
Family Caregiver Alliance https://www.caregiver.org/caregiving	Web portal	Informal caregivers and professionals	Social network, news	Deployed	SP

(continued)

Table 1.1 (continued)

Name	Type of technology	Target group	Endpoints	Status	Countries
Programa Ayuda al Cuidador (PAC), región de Murcia	Telephone and network	Informal caregivers	Help and support when needed	Deployed	SP
SensorMind	Sensors and network	Informal caregivers and relatives	Need of information in real time	Deployed	UK, US
Connect-i http://www.carersmiltonkeynes.org/	Web portal	Informal caregivers, companies and municipalities	Communication, professional support and training	Research (pilot)	UK

Table 1.2 Apps for assistive care

Name of the App	Operative system	Target group	Endpoints	License
Personal Caregiver (imedic8) http://www.personalcaregiver.com/	iOS (iPhone)	Persons needing assistance/Informal an professional caregivers	Medication reminder, alerts, diary, data base, calendar	Personal (Free) Premium ($9.99) Professional ($3.99/user/year)
Elder 411 y Elder 911 (Presto Services) http://www.elder411.net/	iOS (iPhone)	Informal caregivers	Elder 911: management of an emergency situation of the assisted person Elder 411: general information on care, safety	Free
WebMD Mobile (2000) http://www.webmd.com/mobile	iOS(iPhone and iPad) Android	Informal caregivers	Information about medication, treatments, symptoms, etc.	Free
iBiomed (MyRoster) http://www.biomedprofile.com/	iOS (iPhone)	Informal caregivers	Detailed record of the medical information of the person assisted, creation and management of the profiles of the assisted person (medication, tests, diets,…) to track day after day	Free
Pain Care (Ringful Health) http://www.ringfulhealth.com/apps/painfree/	iOS (iPhone) Android	Persons needing assistance	Guide on medical prescriptions and drugs	Free

(continued)

Table 1.2 (continued)

Name of the App	Operative system	Target group	Endpoints	License
iPharmacy (SigmaPhone) http://medconnections.com/	iOS(iPhone and iPad) Android	Informal caregivers	Guide on medical prescriptions and drugs	Free
Pocket First Aid & CPR (Jive Media) http://jive.me/apps/firstaid/	iOS (iPhone) Android	Informal caregivers	Recommendations based on the American Heart Association, information, care	$1.99
Tell My Geo (Iconosys) http://www.iconosys.com/productlist.php?id=249	Android	Informal caregivers	Track the assisted person: disorientation, loss, etc.	Free

deployment of new instances of the system, including the setup and configuration of all required nodes and services. To be sustainable and easily exportable, for example to partners/distributors of the platform in other countries, the process has to be as automated as possible. Once again, APIs at IaaS infrastructure level are key to being able to develop an adequate solution.

Once the system is running, it must be ensured that it has the necessary resources, that the cluster is correctly sized, etc. Any reduction in the cost of cloud services would increase the profit margin. Integration with API of Cloud providers in aspects such as Billing, Customer Relationship Management, etc. in the SaaS layer of the Cloud model; aspects such as access to computing capabilities in the IaaS layer; aspects such as access to the configurability of the deployed instances in a specific environment at PaaS level. In the perspective of the system, the three layers are important although the IaaS layer takes on a special role due to the need to control the computing and storage resources given the needs of the services to be provided.

To guarantee portability/interoperability among different cloud providers, it is necessary to be able to move the supplier platform in order to be independent and have negotiating power, as well as saving costs in the process. The lack of cloud interoperability standards makes this aspect a risk to be managed in an adequate manner. Adequate selection of the provider is of utmost importance. It is not a trivial process and it supposes a great investment and effort. Not all cloud providers offer services in all layers, there are specialists, and a large part of them are companies in the United States with mirrors in Europe. It is very important to be able to detail the Service Level Agreements (SLA) so as not to have risks during exploitation.

The choice of the appropriate cloud model: Public, Private or Mixed is also utmost important. The Mixed will be preferable in order to have greater control of the storage of data, but it is necessary to establish the borders with the suppliers and this will take effort.

Finally yet importantly, privacy problems in addition to other legal frameworks (GDPR and equivalents in the international environment) in multi-geographical environments for the management and preservation of personal data should be considered at a design stage.

1.3.3 Cloud Technologies and Their Application in Ageing

The cloud computing is an information technology paradigm which consists of providing the resources as a service, allowing users to get technological features from the internet on a custom demand. This model matches our needs in the sense that the requirements of computational resources will depend on each algorithm and de data analysed by each of it. The key point for building a cloud computing architecture is the virtualization of the hardware resources, and then, the dynamic adaptation of them to the system needs on each moment according to the demand fluctuations.

To better understand the resulting analysis of this chapter, we have followed the three layer schema of the introduction (Fig. 1.1). This schema provides different levels of functionalities depending on the layer which the system implements. The first layer is named the Software as a Service (SaaS), which represents that the application is offered as a service though the internet. This means that the client does not have to install or maintain the software, it is executed remotely. The second layer is the Platform as a Service (PaaS), which represents the framework where the applications can be developed and executed. The base layer is the Infrastructure as a Service (IaaS) which provides all the hardware resources (servers, network equipment and storage among others).

Two main commercial cloud services which provide these three layers have been compared:

- Amazon Web Services

Amazon Web Services (AWS) is a cloud platform which provides a high variety of functionalities for applications deployment (Fig. 1.2). It offers services classified into: Computational and network Services, Database Services, Storage and Content Delivery Services, Deployment and Management Services and Application Services. The virtual appliance format in AWS is the Amazon Machine Image (AMI), which allows many options with regards to operating systems. Both computational and memory features are configurable, as well as the hardware of the host computer. The cost per hour is determined depending on the instance of the AMI.

Regarding the security, AWS firewall is configured in a deny-all mode by default. Customers have to open desired in-bound ports in the AMI I/O configuration page, only authorized through a X.509 certificate and key. All data is uploaded and downloaded through encrypted communications (https). Another interesting service is the Amazon Simple Storage Service, which allows storing multimedia objects within buckets with different levels of access.

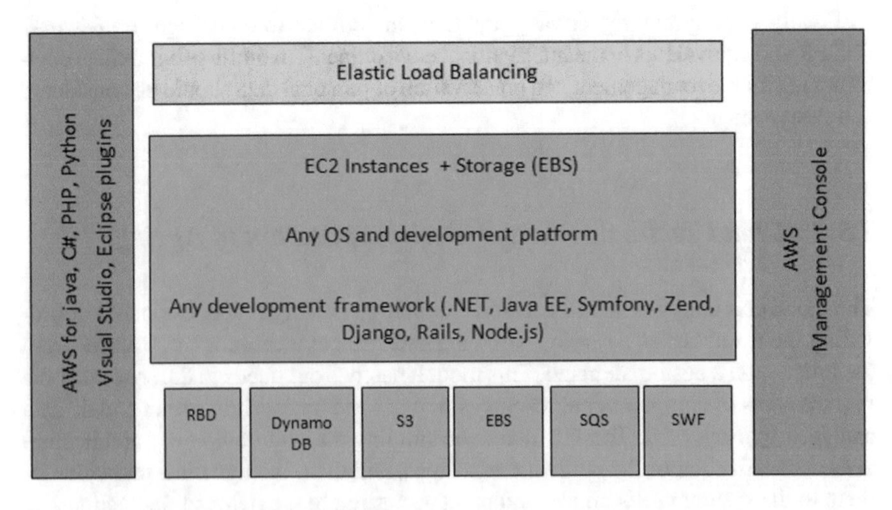

Fig. 1.2 Amazon web services modules

The Amazon Relational Database Service (ARDS) is the engine to set up, operate and configure relational data bases and it is compatible with technologies such MySQL, Oracle and SQLServer. ARDS provides security features for view access, data encryption, SSL connections (only for MySQL deployments), automated backups and replication tools for high load operations (Multi-AZ).

- Microsoft Azure

Azure is Microsoft's cloud infrastructure solution for applications building, deployment and management. It provides services for model execution, data management, networking, business analytics, messaging, development, commerce, media, identity and high performance computing. The virtualization is based on Hyper-V hypervisor and supports Windows, some Linux distributions and Unix Free BSD as operative systems.

Virtual Machine specifications vary from 1 to 8 CPU cores, up to 14 GB RAM, and a bandwidth of 800 Mbps. Each VM is composed by a root partition, which has straight access to hardware resources, and child partitions.

It enables a scripting interface, the PowerShell that allows to control and to automate deployment and management of workloads. Authentication is carried out by SSL and admits own certificates.

It supports development under.NET, Java, PHP, Python and Node.js frameworks, and it is capable of executing any binary file with windows OS compatible. Azure approach at the IaaS management level is to restrict user's access for remote debugging and files, which is restricted by default. However it can be edited using the active directory and allow users to work under specific profiles.

With respect to storage, the Azure Blob tool allows to store any type of files. They can be reached through HTTP/S (no FTP) and based on user's preferences, they can be encrypted. It also allows to reach public material by associating permissions to a container within the blob or assigning it to a Drive. T The Azure Storage Service also supports of shared access signatures which can be used to provide a time-limited token allowing unauthenticated users a time-limited ability to access a container or the blobs in it (Fig. 1.3).

Azure offers a data base engine based on Microsoft SQL Server, the SQL Database. It offers two types of access control: SQL Authentication and a server-side firewall that restricts access by IP address. The SQL Database only supports SQL Server authentication, meaning, users accounts with strong passwords and configured with specific rights. Along with access control SQL Database only allows secure connections via SQL Server's protocol encryption through SSL protocol. It performs real-time I/O encryption and decryption of data and log files. For encryption it uses a database encryption key (DEK), stored in the database boot record for availability during recovery. TDE protects data stored at DB, and enables software developers to encrypt data by using AES and 3DES encryption algorithms without changing existing applications.

AWS and Azure have almost the same functionalities. A cost analysis should be performed before deciding which the most adequate IaaS to be used. However, due

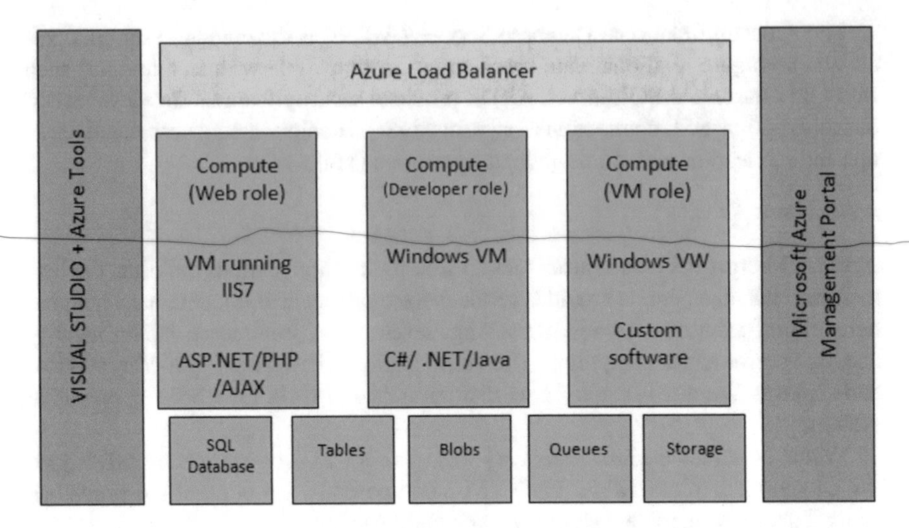

Fig. 1.3 Azure modules

to the support on.NET technologies and the vast amount of plugins for integrating heterogeneous applications, the Azure IaaS seems to be the right choice. It is more accurate in their purpose and then, can offer high efficient performance. However, at the time of migrating to a cloud system this options will be reassessed.

Besides the commercial cloud platforms, there is a European initiative named FI-STAR (https://www.fi-star.eu/about-fi-star.html) which is currently proposing an open stack cloud based framework for the integration of ICT systems, the FI-WARE infrastructure. FI-WARE is currently a project that will provide an open source platform, based upon a series of elements (called Generic Enablers) which offer reusable and commonly shared functions serving multiple areas of use across various sectors, for instance:

Cloud Hosting: Is the fundamental layer which provides the computation, storage and network resources services to be provisioned and managed. The abstraction of software and hardware resources is managed by Generic Enablers (GE) and the Service Management (SM) modules, which interact across them to provide a dynamic environment for system hosting and development.

- IaaS Service Management (SM) GE may invoke APIs of IaaS Data Center Resource Management (DCRM) GE to perform operations on virtualized resources (mainly virtual machines) which comprise the services managed by SM GE.
- Cloud chapter GEs uses Identity Management and Access Control APIs for authentication and authorization purposes.
- IaaS Service Management (SM) GE may use APIs of the Monitoring GE to collect metrics of the underlying resources which comprise the service, to drive service elasticity.

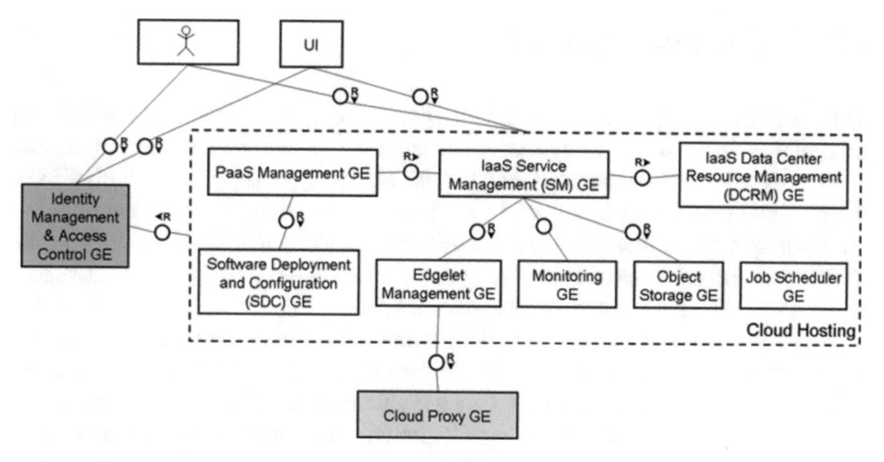

Fig. 1.4 FIWARE Cloud architecture (source FIWARE)

- PaaS Management GE will use IaaS Service Management GE to drive provisioning and auto-scaling of the VMs composing the PaaS software stack.
- PaaS Management GE will use Software Deployment and Configuration (SDC) GE to install and configure the software components running within the individual virtual machine comprising the PaaS environment (Fig. 1.4).

FI-WARE architecture is available to developers, through a working instance named FI-Lab via internet. As an innovative platform, this system may be deployed on it to be a proof-of-concept of healthcare/social-care systems that merges big data repositories, heterogeneous models and algorithms and user interfaces for web/desktop/mobile applications, as we will see in forthcoming sections.

The application of cloud technologies in for supporting informal caregiving are scarce. The most similar context is the Ambient Assisted Living (AAL) domain, which interconnects three different domains of technologies: teleheath, telecare and smart homes. Telehealth technologies are mostly information, communication and sensor technologies that are used to assist people that have medical conditions (e.g. remote patient monitoring systems for chronically ill patients with diabetes or COPD (Chronic Obstructive Pulmonary Disease), electronic pill box, etc.). Telecare technologies are similar set of technologies as used in telehealth but used to assist old people or people with disabilities in social care and to leave independently (e.g. personal emergency response systems, navigation systems, etc.). Smart homes (domotics) technologies are used to automate building controls and to minimize people efforts in maintenance. The AAL domain usually refers to the above mentioned technologies that assist old people (65+) to live independent lives and provide assistance and care either in specialized institutions (social care centres) or in their own residence, but not specifically for assisting informal caregivers.

1.3.4 Opportunities for IoT

There are several opportunities in the Internet of Things (IoT) paradigms which can be exploited in the proposed system and scenario. The main feature of IoT is ability to identify and track technology and assets. All “things” are only identified from the rest, and can belong to a domain of “things”. Therefore, it seems important to have a unique global identification structure that allows to identify classes of objects, individuals and other assets. Technologies such as RF-ID or simply wireless sensor networks (WSN) help to realize this concept [22].

Another feature of the ioT paradigm is the ability to communicate things. By nature “things” in most cases will be simple objects, although there may be complex, and therefore the information they are able to give will be as complex as it is its functionality or its role in society. Communication should comply with certain restrictions: energy efficient technologies, and light (wireless) technologies, which leads us to multi-frequency systems (due to the potential coexistence of many things in one physical place) or frequency management, combinations of different mechanisms (e.g., WiFi and passive RF-ID) depending on the nature of the communication [31].

Finally, the capacity to form networks within the scope of the assisted user, for instance by creating networks with large numbers of independent nodes. IPv6 is a candidate for network management, or wireless technologies in the physical medium such as WiFi, or mobile public communications (probably the objects in this case would be deployed on advanced devices such as smart phones) [40].

1.3.5 Challenges of the Internet of Things (IoT)

The objective of the implementation of this paradigm is to provide the environment where the elderly or assisted person lives from a space with sensors but without addressing their physical space in a tremendous way. That is, the goal is to install as few devices as possible (no more than 5) to give the greatest value to the family member and to the person assisted, but trying to be as unnoticed as possible. It is possible that the informal family caregiver and the person live in the same environment, or with more people. The system will be efficient to the extent that it serves in those cases in which the assisted person is alone or lives alone, which is when there is a lack of attention and when the informal caregiver does not know what is happening.

It is possible that some of the devices to improve its effectiveness have to be carried by the assisted person. A clear example is the system of detection of falls, since an installation without the person having to carry this device becomes a much more expensive facility to develop the same purpose. The device chosen will have to be washable and easily manageable in terms of “removing and put” especially in the case of people with limited abilities in the mobility of limbs such as arms, shoulders …

The objective of the system is not to develop own sensors but identification and the integration in the market, their evaluation and their selection, minimizing the impact of technology obsolescence or lack of distribution in case of exploitation. This approach, aligned with the implementation of standards, will avoid the "vendor lock-in", and will allow us to being able to change suppliers without affecting the final solution. Following the same points defined in the previous section, we will explain the development to be carried out during this project around the home infrastructure.

- Identification technology: devices will be deployed throughout the house with very specific functionalities. Each of them will be uniquely identified and will fulfil a specific function. The things, or sensors, will be totally wireless, without power cable, so that they are easily installable in any place and increase the acceptance of the system by their users.
- Ability to communicate things: the Internet of Things infrastructure by definition makes things connected to the Internet. In this case, in order to lower costs and better control the local network environment, one of the objects in the network will be able to communicate the information to the cloud on behalf of the rest. This particular "thing" is the so-called Gateway, and it also has the capacity to house one or several sensors (transducers) in it. The communication mechanism to the Internet will be flexible and may involve the contracting of a broadband network in the home or maintaining a narrow bandwidth (for example, communicating by pulses).
- Capacity to create networks: a dynamic local network will be created in a mesh form so that any node at least is able to reach another node and at least one node has access to the aforementioned Gateway. One of the reasons that lead us not to implement a pure Internet of Things, that is to say that all the nodes of the network are connected to the Internet is on the one hand the cost, since it is preferable to delegate that aspect to a single device in the network and the second is network coverage. Imagine a real environment where each sensor has a WiFi communication interface and is connected to the Internet, therefore through the home router. If the architectural distribution configuration of the home does not allow any room to have WiFi coverage in that room, it will not be possible to install a sensor. If that instance is the bedroom or bathroom, key environments, the system is unusable. In a configuration with a ZigBee type technology, while another node in the network is reached from that room, the information will be routed appropriately until reaching the Gateway.
- Capacity for self-configuration and, therefore, ability to publish what a "thing" offers and to discover those around it or achievable. In this regard, the network in the home and thanks to the intervention of the Gateway will be self-configuring allowing the addition, elimination or replacement of devices without any penalty in the operation. It is not expected that the family member or the assisted person will be in charge of such management, but the provider company as part of the service.
- Advanced technologies in the "Things" such as:

(a) Embedded intelligence (ability to execute programs): it will be the Gateway that implements the intelligence necessary to transmit the information adequately to the cloud.
(b) Reduced circuit hardware (nanotechnology, small and high-performance hardware architectures, MEMS, energy storage and limited memory in size …) and (e) efficient energy management; they do not foresee development in this area. The only noteworthy aspect is that the sensors will be wireless, to favor the installation in any place even hidden, and it will be necessary to define communication protocols with the Gateway that are efficient in the use of the network from the point of view of saving energy consumption.
(c) Use of M2 M information exchange protocols: between the Gateway and the cloud, M2 M protocols are implemented to provide information in a reliable manner. It is important to use standards at this level, such as the FG M2 M of the International Telecommunication Union, so that the Gateway device to be developed has sales opportunities by itself as a functionally interesting device.
(d) Information modelling and composition of services through the use of semantic technology.
(e) Management of security, identity, and even the reputation of things.

1.3.6 Taking Advantage of the Intelligence

Drools provides support for the runtime management of events, events and rules, so the system will take advantage of the following advantages to allow caregivers to dynamically configure their own intelligent rules through an intuitive interface:

- Adapt dynamically to changes in reality.
- Increase or decrease the number of rules in the system and therefore modify the reactivity of the same.
- Update the accuracy of the rules manually or automatically so that learning occurs from the feedback of the system.

In this aspect, research is key to offer a good number of useful and reliable rules that allow the base to have a good intelligent functioning of the system. The complexity of the rules and the great variability of potential situations to be detected make the result of this research an enormous value that must be protected. Thus, this knowledge will be also susceptible to be protected.

1.3.7 Proposed Architecture

The technological innovation we propose is based on the capacity of the technologies in Cloud Computing to lodge the system functionalities. However, although Cloud technologies have advantages that moved us to consider this strategy, there are also challenges. It is precisely in these challenges in which the system should work intensively with the objective of configuring a viable technology solution from the economic and efficiency point of view, but also from the technical and legal perspective (Fig. 1.5).

As a dynamic platform, the system architecture requirements should be flexible and may depend on the algorithms and rules being executed, the amount of data to be processed or the type of service requested at any time. The high variability of the demanded resources by each home and caregiver may derive in an inefficient use of the hardware and software capabilities of the system, therefore, the design of the architecture has been based on Cloud technologies and SaaS (Software as

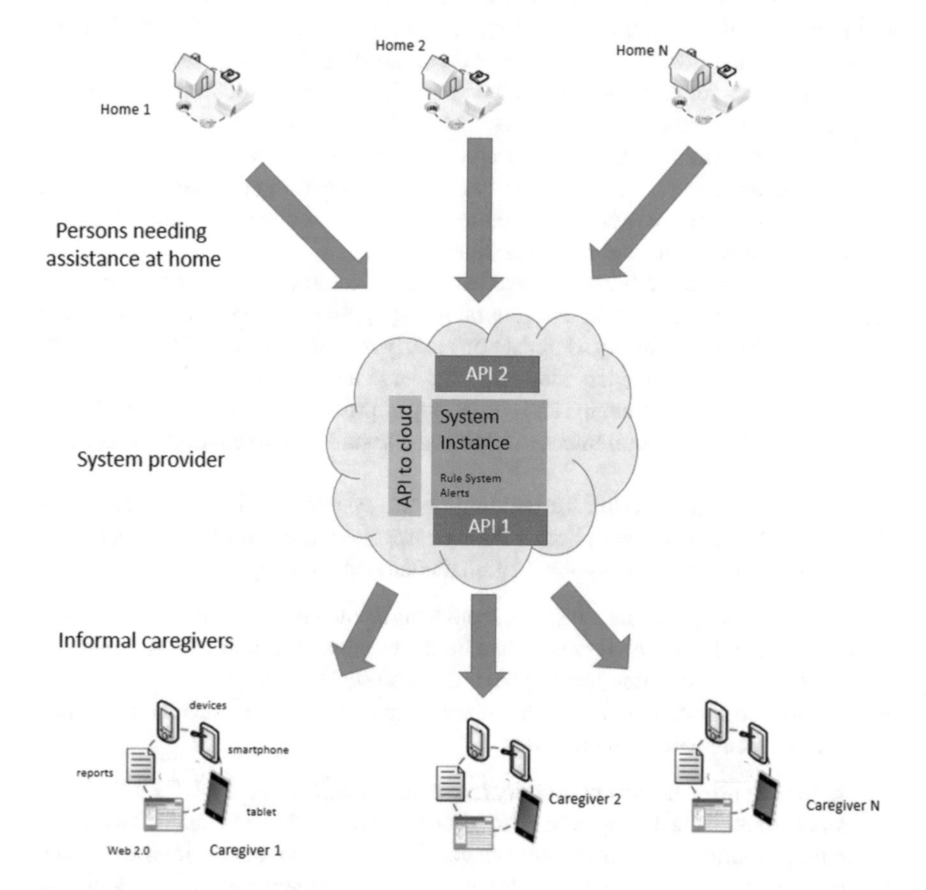

Fig. 1.5 Proposed system architecture

a Service) approach. The architecture is based on three functional modules (data storage, models host and plugin clients) and three operational layers (data access layer, business layer and service providers) that balance the amount of resources demanded by each process within the system.

The operational layers are in charge of communicating and exchanging information between the modules using web services, which finally builds a Service Oriented Architecture (SOA) built over a Cloud Computing infrastructure. This approach has been used by researchers in the clinical context previously [25]. To achieve the service orientation it is needed to ensure an orchestration and choreography for the collaboration of services. The orchestration is based on the cooperation of services in a controlled work flow, whereas, choreography relates to observe the exchange of messages, rules of interaction and agreements between two or more autonomous processes. In a first stage of design, the system architecture was based on three Use Cases, each of them defining a specific work flow which has to do with retrieving data, executing algorithms and displaying the outcome to the user for a decision support. Nevertheless, as the system got mature, each of these phases in the work flow were understood as autonomous processes, bringing to the scene the requirement of design the architecture with both a choreographer and an orchestrator. The key design attributes for orchestration include participant and role definition, variables, properties which enable conversation, fault handlers for exception processing, compensation handlers for error recovery and event handlers to respond to concurrent events with the process itself and a set of activities. Regarding the choreographer, it is required to design the message structure, asynchronous communication, message rules, invocation, events and event handling.

To host a distributed SOA it is needed a specific infrastructure, which may be scalable, able to integrate different programming applications and technologies and furthermore, allow flexibility to update or modify modules within it. The execution of algorithms and data feeders may request a large amount of resources, therefore, performance of data base engine and computing of processes is a critical issue for the system as well as the requirement of enabling a secure way to connect all modules in the proper way.

A quick and easy solution for building up the system could have been a single Application Server (Windows Server 2008 or Apache-tomcat) hosting a data engine (Microsoft SQL Server or MySQL) but that could lead to integration problems:

- Data-type mismatches as migrating from heterogeneous data sources.
- Any modification in the services, data model or messages should be approved by the system administrator, loosing flexibility and scalability.
- Performance limitations in the execution of models and data queries, even they are performed as isolated processes.

These facts bring us the opportunity to use an innovative approach for integrating the system though the deployment of an Infrastructure as a Service (IaaS). This cloud computing solution offers many advantages from the performance, integration and set-up point of view. For the first prototype, the system is going to run over a private cluster of servers, nonetheless, as the system grows up in complexity, it will be moved to a cloud infrastructure. Currently there are several solutions available provided by software companies that may appear very similar. Main differences can be spotted in the cost of their services and the performance, therefore a cost comparison is arranged after a technical description for each cloud service.

1.4 Discussion

In this chapter we have proposed a system architecture focused on the communication requirements of informal caregivers. Whereas current systems are focusing on the integration of wearables and mobile technologies for collecting data in a pervasive way, we focused our proposition in the exploitation of current technologies which may improve the scalability of these systems: Cloud Computing, IoT and smart rules.

From the analysis of the previous applications it is clear that today almost all existing initiatives focus on providing the informal caregiver with 'extra information' about the care of the elderly or dependent person; most of them provide information about the illness, care, treatments or symptoms [13]. Applications such as Personal Caregiver or iBiomed go a step further and they allow to create profiles of the people assisted, managing their daily medication, tests or setting alerts and reminders. However, none of the applications currently available in the market, collects information based on a series of environmental sensors, key aspects of the daily life of a person in a situation of dependency. In addition, in our work we noted that the majority of the consulted applications are available only in English language, being this a great barrier both for the people assisted and for their carers.

Most of the analysed applications do not have a direct technological support to the informal caregiver. They simply establish a support level of documentation and training and never connecting the informal caregiver with the family member, or if he does it indirectly as in the case of the classic telecare [26]. Only the ACTIVAGE solution has functionality similar to that proposed in this chapter, but with the limitations of monolithic systems [12]. None of the solutions incorporates autonomous intelligence and data processing analysis. Only "sensorMind" offers alerts but they are fixed and the informal carer cannot configure them. None of the proposed solutions incorporates an application on smartphone (either on Android or iOS) that complements the web access.

We have presented a solution based on connecting the informal family caregiver directly with the assisted person without any intermediary, but through a private space in the cloud. This system is able to store the information of the sensors and other information of both the caregiver itself and the assisted person. The functionalities offered by this system and the opportunities of the technologies are: (1) Storage and consultation of relevant events through a daily report. (2) Analysis of trends regarding behaviour and respect to the occurrence of certain alerts. (3) Intelligence in the generation of alerts that are truthful and that can be distinguished in several types: informative, important and urgent. (4) Configurable alerts customized by each caregiver for each person assisted, in addition to the inclusion of well-proven general alerts. (5) Availability of the option of a caregiver's diary for the unstructured recording of personal perceptions for personal follow-up. (6) Documentary access to relevant and contrasted sources of information. (7) Access links to the most active social network groups around the figure and the caregiver's problems (Table 1.1) and (8) A communication channel with the provider to report about incidents that may occur in the acquisition infrastructure at home of the person assisted.

Besides, due to the double flow of information the system could be useful to report on new functionalities, designing a space in which the family member can adequately describe their proposals, even being able to make sketches and attach documents.

The proposed system is a solution that aims to evolve from the description made in this document towards an integral solution of care for the informal caregiver of assisted people (elderly, dependent and large dependents). The ability to evolve is embedded in the architecture of the system, an aspect that especially affects the Back-End Subsystem in the Cloud system.

As a result, we have designed an efficient integration of a heterogeneous group of technologies to offer a state-of-the-art technological solution which with the following features: (1) Thanks to cloud computing, the system keeps the efficiency in the management of the deployments for a massive number of users and distributed geographically throughout the world. (2) From the perspective of user's, the system keeps access to information from anywhere and whenever it is needed, for example when the informal caregiver is with the person assisted. (3) The system keeps the simplicity for the user and the initial cost savings for the minimization of equipment in the home thanks to Internet of Things technologies. (4) The system keeps its potential for therapeutic leisure and the ability to access information ubiquitously or to enter information in the system thanks to the integration of applications on Smart Phones and Tablets.

There are several challenges that industry in the AAL domain is facing. First off all, AAL technologies are very often privacy invasive as they often include a sort of user monitoring (data which is collected might include presence information, user's vital signs, environmental data, etc.). Therefore, there are important privacy and confidentiality issues that industry has to solve to ensure wider adoption of the services proposed in this chapter. User studies show an increasing number of end users who identify confidentiality as one of the reasons for their reluctance to adopt these technologies [7]. Secondly, technology must be intuitive and easy to use (functionality such as remote configuration and exception management are very important) as they are mostly aimed at aged people (65+). Vendors should be capable of integrating and extending their products and services and support interworking with devices from other vendors, as partial and closed box solutions are not desired and are more costly.

Another challenge os the proposed architecture is the initial set-up costs, which is another issue the industry has to confront to. Finally, low awareness by the user community and unclear reimbursement models are other limiting factors for the adoption of supporting technologies for caregivers.

1.5 Conclusion

The quality of life of any person, young or old, depends to a large extent on efficiency, comfort and the place he or she calls "home". People with disabilities have specific requirements regarding their home environment and its functionalities. The paradigm proposed in this chapter integrates existing services and solutions into a scalable system in which current and future technologies can contribute to support both the assisted persona and the caregiver.

These type of technologies, in the core of Ambient Intelligence and Ambient Assisted Living, when focused on the elderly and/or dependent user may help people with functional diversity and the increasingly large group of elderly people to live longer in the place they like the most, while ensuring high safety, including health surveillance and health care functions. We have evolved the concept of ubiquitous computing to improve the use of sensors [1] and computation nodes (massively distributed throughout the physical space but invisible to people) so that the technology is transparent to the user.

The application of the combined Cloud computing frameworks and Internet of Things in monitoring devices and other services, allows the person in question to remain in their domestic environment. Receiving social services and medical support through new channels that integrate into the home contributes to improving the quality of life of many elderly and disabled people. Future directions should investigate in the integration of these services with Social Care trusts from municipalities and Electronic Health Records. Moreover, cost-effectiveness studies should be designed to test the capability of these systems to prevent loneliness, support an active ageing and mitigate the burden of caregivers.

References

1. Bayo-Monton, J.-L., Martinez-Millana, A., Han, W., Fernandez-Llatas, C., Sun, Y., Traver, V.: Wearable sensors integrated with Internet of Things for advancing eHealth care. Sensors **18**(6) (2018)
2. Bousquet, J., Kuh, D., Bewick, M., Standberg, T., Farrell, J., Pengelly, R., Zins, M., et al.: Operational definition of Active and Healthy Ageing (AHA): a conceptual framework. J. Nutr. Health Aging **19**(9), 955–960 (2015). https://doi.org/10.1007/s12603-015-0589-6
3. Bratteteig, T., Wagner, I.: Moving healthcare to the home: the work to make homecare work. In: Bertelsen, O.W., Ciolfi, L., Grasso, M.A., Papadopoulos, G.A. (eds.): ECSCW 2013: Proceedings of the 13th European Conference on Computer Supported Cooperative Work, 21–25 September 2013, Paphos, Cyprus, pp. 143–162. Springer, London (2013)
4. Chang, F., Östlund, B.: Perspectives of older adults and informal caregivers on information visualization for smart home monitoring systems: a critical review. In: Bagnara, S., Tartaglia, R., Albolino, S., Alexander, T., Fujita, Y. (eds.) Proceedings of the 20th Congress of the International Ergonomics Association (IEA 2018), pp. 681–690. Springer International Publishing, Cham (2019)
5. Chen, W.-L., Chen, L.-B., Chang, W.-J., Tang, J.-J.: An IoT-based elderly behavioral difference warning system. In: 2018 IEEE International Conference on Applied System Invention (ICASI), pp. 308–309 (2018). https://doi.org/10.1109/ICASI.2018.8394594

6. Covassin, N., Singh, P.: Sleep duration and cardiovascular disease risk: epidemiologic and experimental evidence. Sleep Med. Clin. **11**(1), 81–89 (2016). https://doi.org/10.1016/j.jsmc.2015.10.007
7. Dehling, T., Gao, F., Schneider, S., Sunyaev, A.: Exploring the far side of mobile health: information security and privacy of mobile health apps on iOS and Android. JMIR MHealth UHealth **3**(1), e8 (2015). https://doi.org/10.2196/mhealth.3672
8. Dishman, E., Matthews, J., Dunbar-Jacob, J.: Everyday Health: Technology for Adaptive Aging. Technology for Adaptive Aging (2004). Retrieved from http://www.ncbi.nlm.nih.gov/books/NBK97353/
9. Fehling, C., Leymann, F., Retter, R., Schupeck, W., Arbitter, P.: Cloud Computing Patterns Fundamentals to Design, Build, and Manage Cloud Applications. Springer, Heidelberg (2014). https://doi.org/10.1007/978-3-7091-1568-8
10. Fernández-Caballero, A., Latorre, J.M., Pastor, J.M., Fernández-Sotos, A.: Improvement of the elderly quality of life and care through smart emotion regulation. In: Pecchia, L., Chen, L.L., Nugent, C., Bravo, J. (eds.) Ambient Assisted Living and Daily Activities, pp. 348–355. Springer International Publishing, Cham (2014)
11. Fernandez, E.B., Monge, R., Hashizume, K.: Building a security reference architecture for cloud systems. Requirements Eng. **21**(2), 225–249 (2016). https://doi.org/10.1007/s00766-014-0218-7
12. Fico, G., Montalva, J.-B., Medrano, A., Liappas, N., Mata-Díaz, A., Cea, G., Arredondo, M.T.: Co-creating with consumers and stakeholders to understand the benefit of Internet of Things in smart living environments for ageing well: the approach adopted in the Madrid deployment site of the ACTIVAGE large scale pilot. In: Eskola, H., Väisänen, O., Viik, J., Hyttinen, J. (eds.) EMBEC & NBC 2017, pp. 1089–1092. Springer, Singapore (2018)
13. Godwin, K.M., Mills, W.L., Anderson, J.A., Kunik, M.E.: Technology-driven interventions for caregivers of persons with dementia: a systematic review. Am. J. Alzheimer's Dis. Other Dementias® **28**(3), 216–222 (2013). https://doi.org/10.1177/1533317513481091
14. Gubbi, J., Buyya, R., Marusic, S., Palaniswami, M.: Internet of Things (IoT): a vision, architectural elements, and future directions. Future Gener. Comput. Syst. **29**(7), 1645–1660 (2013). https://doi.org/10.1016/j.future.2013.01.010
15. Hagerman, H., Högberg, H., Skytt, B., Wadensten, B., Engström, M.: Empowerment and performance of managers and subordinates in elderly care: a longitudinal and multilevel study. J. Nurs. Manage. **25**(8), 647–656 (2017). https://doi.org/10.1111/jonm.12504
16. Hashizume, K., Rosado, D.G., Fernández-Medina, E., Fernandez, E.B.: An analysis of security issues for cloud computing. J. Internet Serv. Appl. **4**(1), 1–13 (2013). https://doi.org/10.1186/1869-0238-4-5
17. Hopgood, A.A.: Intelligent Systems for Engineers and Scientists. CRC press (2016)
18. Hwang, A., Truong, K., Mihailidis, A.: Using participatory design to determine the needs of informal caregivers for smart home user interfaces. In: Proceedings of the 6th International Conference on Pervasive Computing Technologies for Healthcare, pp. 41–48 (2012). https://doi.org/10.4108/icst.pervasivehealth.2012.248671
19. Ji, Z., Ganchev, I., O'Droma, M., Zhao, L., Zhang, X.: A cloud-based car parking middleware for IoT-based smart cities: design and implementation. Sensors (Switzerland) **14**(12), 22372–22393 (2014). https://doi.org/10.3390/s141222372
20. Kluzer, S., Redecker, C., Centeno, C.: Long-term Care Challenges in an Ageing Society: The Role of ICT and Migrants. Results from a Study on England, Germany, Italy and Spain. European Commission JRC/Institute for Prospective Technological Studies (2010)
21. Krutz, R.L., Vines, R.D.: Cloud Security: A Comprehensive Guide to Secure Cloud Computing. Wiley Publishing (2010)
22. Kumar, N.S., Vuayalakshmi, B., Prarthana, R.J., Shankar, A.: IOT based smart garbage alert system using Arduino UNO. In: IEEE Region 10 Annual International Conference, Proceedings/TENCON, pp. 1028–1034 (2017). https://doi.org/10.1109/TENCON.2016.7848162
23. Larcher, S., Benhamou, P.Y., Pépin, J.L., Borel, A.L.: Sleep habits and diabetes. Diabetes Metab. **41**(4), 263–271 (2015). https://doi.org/10.1016/j.diabet.2014.12.004

24. Manvi, S.S., Krishna Shyam, G.: Resource management for Infrastructure as a Service (IaaS) in cloud computing: a survey. J. Netw. Comput. Appl. **41**(1), 424–440 (2014). https://doi.org/10.1016/j.jnca.2013.10.004
25. Martinez-Millana, A., Bayo-Monton, J.-L., Argente-Pla, M., Fernandez-Llatas, C., Merino-Torres, J., Traver-Salcedo, V.: Integration of distributed services and hybrid models based on process choreography to predict and detect type 2 diabetes. Sensors **18**(1), 79 (2017). https://doi.org/10.3390/s18010079
26. Martinez-Millana, A., Fico, G., Fernández-Llatas, C., Traver, V.: Performance assessment of a closed-loop system for diabetes management. Med. Biol. Eng. Comput. **53**(12), 1295–1303 (2015). https://doi.org/10.1007/s11517-015-1245-3
27. Natis, Y.: The PaaS Road Map (2011). Retrieved from https://www.gartner.com/doc/1521622/paas-road-map-continent-emerging
28. Oliver, D., Foot, C., Humphries, R.: Making our Health and Care Systems Fit for an Ageing Population. King's Fund (2014)
29. Ortman, J.M., Velkoff, V.A., Hogan, H.: An Aging Nation: The Older Population in the United States. Population Estimates and Projections. Current Population Reports. Current Population Reports, 1964 (2014). https://doi.org/10.1016/j.jaging.2004.02.002
30. Pahl, C.: Containerization and the PaaS Cloud. IEEE Cloud Comput. **2**(3), 24–31 (2015). https://doi.org/10.1109/MCC.2015.51
31. Papapostolou, A., Chaouchi, H.: Exploiting multi-modality and diversity for localization enhancement: WiFi & RFID usecase. In: 2009 IEEE 20th International Symposium on Personal, Indoor and Mobile Radio Communications, pp. 1903–1907 (2009). https://doi.org/10.1109/PIMRC.2009.5449853
32. Pham, V.V.H., Liu, X., Zheng, X., Fu, M., Deshpande, S.V., Xia, W., Abdelrazek, M., et al.: PaaS—Black or white: an investigation into software development model for building retail industry SaaS. In: Proceedings—2017 IEEE/ACM 39th International Conference on Software Engineering Companion, ICSE-C 2017, pp. 285–287 (2017). https://doi.org/10.1109/ICSE-C.2017.57
33. Richters, A., Olde Rikkert, M.G., van Exel, N.J., Melis, R.J., van der Marck, M.A.: Perseverance time of informal caregivers for Institutionalized elderly: construct validity and test-retest reliability of a single-question instrument. J. Am. Med. Directors Assoc. **17**(8), 761–762 (2016). https://doi.org/10.1016/j.jamda.2016.05.001
34. Stavrinides, G.L., Karatza, H.D.: A cost-effective and QoS-aware approach to scheduling real-time workflow applications in PaaS and SaaS clouds. In: Proceedings—2015 International Conference on Future Internet of Things and Cloud, FiCloud 2015 and 2015 International Conference on Open and Big Data, OBD 2015, pp. 231–239 (2015). https://doi.org/10.1109/FiCloud.2015.93
35. Tang, P.C., Smith, M.D., Adler-Milstein, J., Delbanco, T., Downs, S.J., Mallya, G.G., Sands, D.Z., et al.: The democratization of health care: a vital direction for health and health care. NAM Perspect. **6**(9) (2016). https://doi.org/10.31478/201609s
36. Tarricone, R., Tsouros, A.D.: Home Care in Europe: The Solid Facts. WHO Regional Office Europe (2008)
37. Truong, H.-L., Dustdar, S.: Principles for engineering IoT cloud systems. IEEE Cloud Comput. **2**(2), 68–76 (2015). https://doi.org/10.1109/mcc.2015.23
38. Valero, E., Adán, A., Cerrada, C.: Evolution of RFID applications in construction: a literature review. Sensors (Switzerland) **15**(7), 15988–16008 (2015). https://doi.org/10.3390/s150715988
39. Wan, Z., Liu, J., Deng, R.H.: HASBE: a hierarchical attribute-based solution for flexible and scalable access control in cloud computing. IEEE Trans. Inf. Forensics Secur. **7**(2), 743–754 (2012). https://doi.org/10.1109/TIFS.2011.2172209
40. Yeh, S., Talwar, S., Wu, G., Himayat, N., Johnsson, K.: Capacity and coverage enhancement in heterogeneous networks. IEEE Wireless Commun. **18**(3), 32–38 (2011). https://doi.org/10.1109/MWC.2011.5876498
41. Zhao, F., Gaw, S. D., Bender, N., Levy, D.T.: Exploring cloud computing adoptions in public sectors: a case study. GSTF J. Comput. (JoC) **3**(1) (2018)

Additional Reading Section (Resource List)

42. Banafa, A.: Secure and Smart Internet of Things (IoT), 1st edn. River Publishers Series in Information Science and Technology (2018)
43. Bloomberg, J.: The Agile Architecture Revolution: How Cloud Computing, REST-Based SOA, and Mobile Computing Are Changing Enterprise IT, Wiley (2013). Retrieved from https://books.google.es/books?id=RU8aYsSDukwC
44. Chast, R.: Can't We Talk about Something More Pleasant?: A Memoir. Bloomsbury Publishing (2014). Retrieved from https://books.google.es/books?id=VKZwCwAAQBAJ
45. Glasby, J.: Understanding Health and Social Care. Policy Press (2007). Retrieved from https://books.google.es/books?id=ZFG-lwEACAAJ
46. Griffiths, D.: Head First Android Development: A Brain-Friendly Guide. O'Reilly Media, Incorporated (2017). Retrieved from https://books.google.es/books?id=SgNdMQAACAAJ
47. Gross, J.: A Bittersweet Season: Caring for Our Aging Parents– and Ourselves. Vintage Books (2012). Retrieved from https://books.google.es/books?id=ZbyebjD6DJUC
48. Harris, J., White, V.: A Dictionary of Social Work and Social Care. Oxford University Press (2018). Retrieved from https://books.google.es/books?id=53CvswEACAAJ
49. Hawranik, P.G., Strain, L.A.: Health of Informal Caregivers : Effects of Gender, Employment, and Use of Home Care Services (2000)
50. Horton, J.: Android Programming for Beginners: Learn All the Java and Android Skills You Need to Start Making Powerful Mobile Applications. Packt Publishing (2015). Retrieved from https://books.google.es/books?id=wByfjwEACAAJ
51. Horton, J.: Learning Java by Building Android Games (n.d.): Retrieved from https://books.google.es/books?id=tA8FCAAAQBAJ
52. Hugos, M. H., Hulitzky, D.: Business in the Cloud: What Every Business Needs to Know About Cloud Computing. Wiley (2010). Retrieved from https://books.google.es/books?id=4GsPIW5TOQYC
53. Kavis, M.J. Architecting the Cloud: Design Decisions for Cloud Computing Service Models (SaaS, PaaS, and IaaS). Wiley (2014). Retrieved from https://books.google.es/books?id=YjIWBAAAQBAJ
54. Mark Hung: Leading the IoT (2017). Retrieved from https://www.gartner.com/imagesrv/books/iot/iotEbook_digital.pdf
55. Meier, R.: Professional Android 4 Application Development. Wiley (2012). Retrieved from https://books.google.es/books?id=48bnSNs_h9sC
56. Rafaels, R.: Cloud Computing: From Beginning to End. CreateSpace Independent Publishing Platform (2015). Retrieved from https://books.google.es/books?id=dGFWrgEACAAJ
57. Rose, K., Eldridge, S., Lyman, C.: The internet of things: an overview—understanding the issues and challenges of a more connected world. Internet Society (October), 53 (2015). https://doi.org/10.5480/1536-5026-34.1.63
58. Shukla, A., Chaturvedi, S., Simmhan, Y.: RIoTBench: A Real-time IoT Benchmark for Distributed Stream Processing Platforms (2017). https://doi.org/10.1002/cpe.4257
59. Trifa, D.G.V.: Building the Web of Things (1st Editio). Manning (2016)
60. Weinman, J.: Cloudonomics: The Business Value of Cloud Computing. (Wiley, Ed.) (2012)

Chapter 2
Assistive Technologies to Support Communication with Neuro-motor Disabled Patients

Cristian Rotariu, Radu Gabriel Bozomitu and Hariton Costin

Abstract The paper describes the evolution in design and implementation of a new complex electro-informatics assistive medical system used for the communication with neuromotor disabled patients. These patients are severe handicapped persons who cannot interfere with real world by normal ways like speech, writing or language signs. Fortunately, these persons can see or can hear and they understand the meaning of different messages. The developed medical system is made by the following main functional components: the patient equipment, server, and caretaker device, connected together by using Internet available infrastructure. The bidirectional communication is performed by using keywords technology or ideograms presented on the patient's screen provided by an application running at the server and accessed by the patient through a web browser. Patient needs are performed by using for interfacing with a software application by using different types of sensors including switch-type sensors (tactile, pressure, strain gauge) or eye tracking devices, all of these adapted to the patient's physical condition. The medical system has also a secondary function, the remote monitoring of few vital physiological parameters needed for patient treatment and rehabilitation, based on wireless and wearable sensor network, that share the same hardware platform. The application running on caretaker device displays the patient's needs together with the values of the selected physiological parameters in real time. The medical devices are responsible for physiological parameters acquiring, preprocessing and forwarding into the

C. Rotariu (✉) · H. Costin
Faculty of Medical Bioengineering, Grigore T Popa University of Medicine and Pharmacy, Iasi, Romania
e-mail: cristian.rotariu@umfiasi.ro

H. Costin
e-mail: hariton.costin@umfiasi.ro

R. G. Bozomitu
Faculty of Electronics, Telecommunications and Information Technology, Gheorghe Asachi Technical University, Iasi, Romania
e-mail: bozomitu@etti.tuiasi.ro

H. Costin
Institute of Computer Science of Romanian Academy, Iași Branch, Iasi, Romania

H. Costin et al. (eds.), *Recent Advances in Intelligent Assistive Technologies: Paradigms and Applications*, Intelligent Systems Reference Library 170,
https://doi.org/10.1007/978-3-030-30817-9_2

server. The server performs the parameters processing and computing the medical condition of the patient and also has the ability of sending alerts to the caretaker when one or more parameters are outside the normal limits. The application running on server manages the system databases, the communication protocols between patient and caretaker devices, records and processes the remote monitored data values and alarms the caretaker devices in emergency situations. The prototype of the developed medical system has been successfully within Clinic of "Geriatrics and Gerontology" at "Dr. C. I. Parhon" Hospital of Iaşi, Romania. The assistive solution proposed can be used in hospitals, care and treatment centers, nursing homes and also for patients treated in outpatient settings. The system contributes to increasing the level of care and treatment for neuromotor disabled patients, lowering costs in health care system.

Keywords Medical rehabilitation · Human-computer interfaces · Neuro-motor disabled patients · Assistive devices

2.1 Introduction

The number of people with neuro-motor disabilities is increasing every year. There are more than two million of disabled patients the USA (approx. 1% of the entire population) and about the same situation exists also in Europe. The exact number of the disabled patients that fit in the category addressed by the proposed medical system is not known, as nobody thought that they form a separate category.

As results from the thematic research areas, at present, internationally efforts are made to improve the quality of care for the people with disabilities. The intensification of the efforts to improve the patients' care is accompanied by a limitation of expenses at reasonable levels. In order to satisfy these two opposing requirements, one solution is for caretakers to use on a larger scale various equipment, based on IT&C technologies, which have become more complex and sophisticated. Among these, the decreasing price of the embedded systems has a positive effect in design and implementation of medical systems.

By allowing the disabled people to perform tasks impossible in other circumstances, the assistive technology improves the way of interacting with the required technology for fulfilling that task [1, 2].

The strategy "Europe 2020" elaborated in August 2010 at Brussels by the EU proposed five ambitious objectives—employment, encouragement of innovation, ensuring education for all citizens, social inclusion for all society members and solutions to all climate aspects, in the context of energy consumption increase; all aims are to be fulfilled by 2020 [3]. Among the actions included in the Strategy Europe 2020 there is also "sustainable healthcare and support supply based on information and communications technology (ICT) for a decent and autonomous life". According to the documents adopted, COM (2007)860 and SEC (2009)1198, the initiative concerning pilot projects in e-Health "will promote standardization, interoperability testing and certification of the registers and electronic equipment in the field of health". In the same strategy, Europe 2020, the European Committee concludes that "the AAL technologies (Ambient Assisted Living) bring ICT to reach all" and thus "there will be taken new actions to put ICT in the service of the most vulnerable of the society members". The aims of this action will lead to "AAL program consolidation" in order to allow the disabled or chronically ill people to benefit from an autonomous and decent life, and to be active in society [3].

Remote monitoring of physiological parameters could enable hospitals or treatment centers to deliver medical services at remote locations to take care of patients in different locations. It allows the use of remote connected medical devices for diagnosis and treatment of patients in different locations by the health care professionals.

The use of internet-embedded devices to undertake remote consultations with clinicians or other health care professionals, to access personal medical records and to connect patients with family members, has already become commercially available. Specialized medical software, data storage servers, and medical devices capable of medical data collection, storage and transmission are available on the market as parts components of the telemedicine infrastructure. Such systems have already been developed in the USA and Europe.

Nowadays, the home-based technologies that use the Internet of Things concept support disabled patients to manage their long-term medical conditions. They become widely used as an evidence of their increased impact and decreased unit costs. There is the potential for many disabled patients to access and use such technologies in this way.

2.2 Background/State-of-the Art

Concerning the current stage of caring methods for disabled people which assures physiological needs (feeding, defecation, micturition etc.) but also medical investigations and their observation. However, the communication systems designed until now to communicate with these people often ensure only one-way communication, hardly serving both patient and supervisor [4–29].

Communication between patients suffering from brain diseases including neurological, ophthalmic vestibular or paralysis, spinal cord injuries etc., that have cognitive function but are not able to speak or to perform hand movements, represents an important issue of their connection with the outside world that need to be solved.

In Romania, there have been long time preoccupations with developing applications for communicating with and remote monitoring of the people with different diseases, in some research centers in Bucharest, Iaşi, Timişoara and Cluj. Projects were developed at "Gheorghe Asachi" Technical University of Iaşi (TUIasi) and "Grigore T. Popa" University of Medicine and Pharmacy of Iaşi (UMF Iaşi). Mention can be made of the project in telemedicine named TELMES ("Multimedia platform for complex medical teleservices"), a project finalized in 2007 and considered the first pilot-project of telemedicine in Romania. A project of assistive technology, developed by TUIasi in collaboration with UMF Iaşi was: TELPROT ("Communication system with people with major neuromotor disability", finalized in 2009) [23]. A more recent remote monitoring project, developed and implemented by UMF Iaşi together with TUIasi was TELEMON ("Integrated real time telemonitoring system for patients and older persons", finalized in 2010) [30]. Another project of assistive technology was finalized by TUIasi in 2011 (ASISTSYS, "Integrated System of Assistance for Patients with Severe Neuromotor Affections") [22]. All these projects have been financed by Romanian Ministry of Research and Innovation and provided a strong relationship between a technical and medical univesities. Work on the projects mentioned above led to notable results, materialized in prototypes and many published papers. Still, there is no operating system in this country for communicating with and remote monitoring the severely neuromotor disabled people, as all the above-mentioned projects produced only prototypes. The only system with operating functions which somehow resembles our proposed system, but was much more specialized and simpler, was built by our team and named TELPROT [23] and ASISTSYS [22].

These systems ensure simple communication, used only to satisfy the immediate needs of the patients.

The experiments made with severely disabled patients in ASISTSYS [22] have emphasized the need to implement a more complex system, designed to give patients the possibility to communicate bi-directionally with the physicians, supervisors and family, by using complex phrases. Considering the need for efficient care and the existing hardware availabilities for communication, the implementation in the same system of remote monitoring equipment for several important physiologic parameters will certainly prove to be useful.

At international level, remote monitoring of disabled patients, especially patients suffering from neuromotor diseases, has become a widely accepted method for evaluation the medical conditions, for diagnosis and rehabilitation, as proven by numerous studies and projects, many of them finalized and/or still in progress [31–41]. Examples of these are HEARTS ("Health Early Alarm Recognition and Telemonitoring System") [33], U-R-SAFE ("Universal Remote Signal Acquisition For health") [31], EPI-MEDICS ("Enhanced Personal, Intelligent and Mobile System for Early Detection and Interpretation of Cardiological Syndromes") [35] and CodeBlue (as a reference project made at Harvard University) [34].

At national level, efforts in the field of remote monitoring, which is a technology of medical interest, are still in the early stage [42–46]. Examples of projects are: CardioNET ("Integrated system for continuous surveillance in e-health intelligent network of the patients with cardiac diseases) [47], TELEASIS ("Complex system, on NGN support—Next Generation Networking—for home senior teleassistance") [48], MEDCARE ("Monitoring system for cardiac diseases, for receiving, transmission by Internet and real time analysis of the electrocardiographic signal") [49], TELMES ("Multimedia platform for complex medical teleservices") [50] and TELEMON ("Integrated real time telemonitoring system for patients and older persons") [43].

Despite all efforts, even at international level, the remote monitoring is still not widely used in the health care systems, many aspects being still under research in order to find the best solutions. One of these problems is represented by the costs, still too high.

There are many healthcare medical systems available on the market that help patients to communicate with clinicians, but these systems are complex and expensive and they are very difficult to be used by disabled persons who may have difficulty with mobility and communication.

For people suffering from hearing diseases a lot of low cost hearing aids are available, for sight-impaired people there are a lot of Braille-type systems, and for the patients with not so severe motor disabilities there are switching type devices, mini keyboards, many of them provided by low range radio transmission [1, 3].

Controlling virtual keyboards with the aid of assistive technologies is possible in some circumstance for the neuromotor disabled patients by triggering virtual switches or by moving a cursor on the screen with their eyeballs [51].

Assistive technologies (AT) are also designed to improve the functional capabilities of people with disabilities allowing them to perform ordinary tasks or to interact with objects during daily activities. In the last years, due to the advances in the IT&C technology and in the field of electronics applied in medical devices, there is a need to develop new assistive devices, suitable to be integrated in more complex systems. These devices have to be adapted for different types of disabilities and to include electronic modules either for diagnostic purposes or assisting disabled patients in their environments [52, 53].

Nowadays, the interest of many researchers working in field of AT is focused on the controlling of the objects by moving the patient eyeballs. The detection of movement of the eyeball are performed by several non-invasive techniques including Infrared Oculography (IROG) [54–56] (measuring the reflected infrared light from a source focused at the eye), Video Oculography (VOG) [57–60] (recording the movement of the eyes by using a video camera) or Electrooculography (EOG) [61–63] (sensing uses the bio-potential produced around eyes due to eyeball motion). Among these technologies, the EOG is a cost effective technique and it is used in many applications for controlling wheelchairs [64], computer virtual keyboards [65, 66], or televisions [67].

Communication with the neuro-motor disabled people is done in various ways, depending on the seriousness of the disability, and the welfare level of the person or system. For less severe disabilities, where mobility is still possible and the patient can execute controlled movements, various communication systems are available: ringing devices (one, two or three) with various tonalities, keyboards—from simple to sophisticated—, with or without display (see, for instance, the products of EnableMart company [1]). These systems make the work of the caretaker easier: the latter needs not to be present near the patient all the time. On the other hand, ringing devices are alarming devices, not communicating ones. They can transmit 1–3 simple pieces of information and can be used only by those patients who can execute well-controlled movements. They are not useful for those patients who cannot move their limbs or speak. For these patients, the approach needs to be different:

(a) For rich people specialized caretakers can be employed assisted by robots—computerized electromechanical devices (e.g. the actor Christopher Reeve). In the developed countries, there are specialized clinics and care units for such patients. The costs are huge and the access is limited to the rich.
Even when 3–4 people take care of one patient, there may be problems of communication, as caretakers may not be highly qualified or dedicated. For this season, communication with disabled patients continues to be a topic of research.
(b) For people with modest material possibilities in both developed countries and the great majority of the developing countries, such as Romania, hospital care is provided during hospitalization periods according to necessities. Home care is provided by relatives, sometimes helped by hired people, with no qualifications.

2.3 Communication System for People with Major Neuro-locomotor Disability—TELPROT Project

The first attempt in building a communication system with patient s having neuro-locomotor disability is represented by TELPROT project. The project started in 2007 promoted research connected with designing and implementing an electronic medical system used for the communication with people having deprived communication capabilities due to major neuro-locomotor disabilities. These people can hear or see, but unfortunately they are not able to communicate by using speech, write or sign language. Many of them can produce uncontrolled muscular contractions or inarticulate sounds, can make some movements but without any possibility to communicate. In hospitals, medical care and treatment units, or at their homes they need specialized care services care attendance. Finding out the necessities of these people, either urgent (urination, defecation) or less urgent (pain, thirst, hunger) is a big problem that implies a continuous effort of the attendants or caretakers, who cannot leave their bed side.

The main objective of TELPROT project was the design and development of a medical system made up of two units (sub-systems): one for the patient and one for the caretaker. The patient unit includes a movement sensor connected to a control unit having a display and a receiver with a loudspeaker. Mainly, the whole system operates in the following modes: at patient's request (movement of finger, inspiration, blinking eyes) the video display of several key phrases (such as "I'm thirsty", "I feel pain", "I want to use the toilet"), in succession. Each sentence is repeated 2 or 3 times, separated by breaks. Hearing or seeing the phrase needed, the patient reacts (in the same way as the request) and the selected phrase is transmitted wirelessly to the caretaker, who finds out what the patient needs and whether this requires urgent or less urgent action. The product has been designed to be used in hospitals, medical care and rehabilitation units, patient's homes, or asylums, etc.

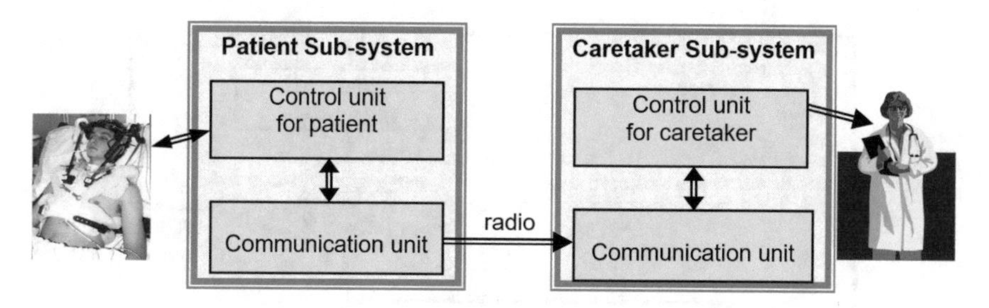

Fig. 2.1 The functional sub-systems used for patient—caretaker communication

This system illustrates integrated technology used for implementing an individual interface that interconnects people with temporary or permanent disabilities to services and resources.

The TELPROT system is made by two main functional components that share the same functional units: the patient sub-system and the caretaker sub-system. The patient unit is composed by a control unit used to acquire and process the information coming from the patient and a communication unit, while the caretaker unit contains units with similar functionalities, but designated to be used by caretaker (Fig. 2.1).

Figure 2.1 briefly shows the functional structure and components of the of the patient sub-system:

(1) a display module used for visual representation the keywords;
(2) a loudspeaker for the audible form of the keywords displayed;
(3) an appropriate switch for the patient illness, controlled by the patient;
(4) a control unit used to allow successive keywords to be displayed and heard and to permit the selection of keywords by using a switch;
(5) a communication unit using radio waves.

Figure 2.2 presents the principles of operations in TELPROT system. In its normal state the patent sub-system is in standby mode and it is activated by the patient through the appropriate switch.

During this time, the keywords set are activated and the patient can see them on the control unit display and hear them on the loudspeaker. Each keyword is active on the screen a short period of time that can be configured by the caretaker at the first use of the system with the patient. If the patient wants to transmit a keyword he/she must activate the switch (putting it in the ON state), wait to see and hear the question "Transmission?", and give a positive answer by activating the switch again.

Appropriate switches (sensors and/or actuators) are available on the market in a large variety of types and variants, all of them containing at least two positions: one active and one inactive. The differences regard the principles of operation, shapes and dimensions are presented in Fig. 2.3.

The keywords sets have been clearly spoken, recorded and digitized and stored into a local non-volatile memory. For displaying the keywords to the patients, a

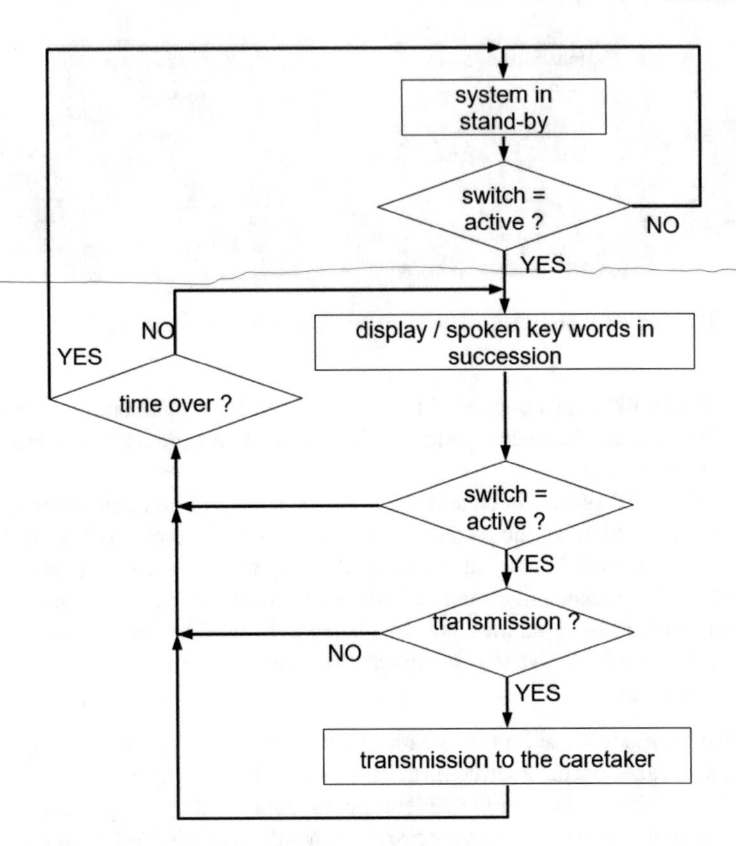

Fig. 2.2 Flowchart of operations in TELPROT system

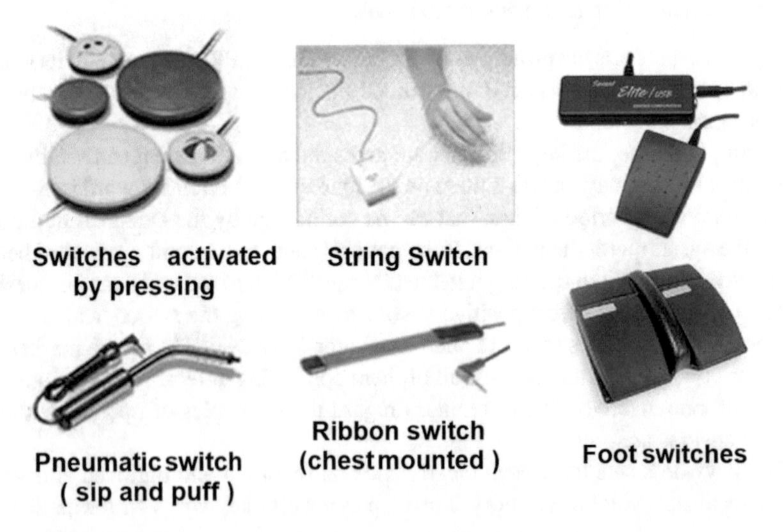

Fig. 2.3 Different switches (sensors and actuators) for disabled people

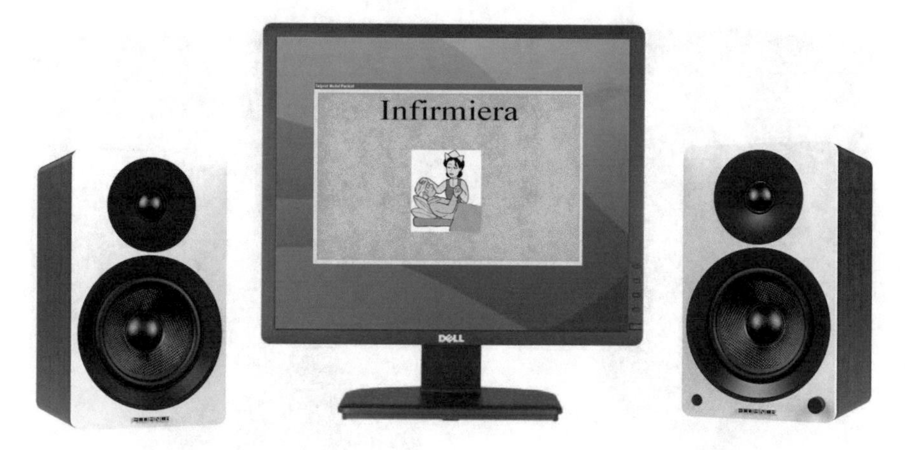

Fig. 2.4 Graphical user interface (patient sub-system) (Romanian word for "nurse")

standard 15-in. LCD with a resolution of 1024 × 768 has been used. The radio link uses communication modules operating in the Short-range device band, which does not need licensing.

The software application developed for the TELPROT system use the client-server model. The client module contains a software application, in form of a Graphical User Interface (GUI–Fig. 2.4), that interacts with the patient in the following modes:

(1) by displaying the keywords set and associated images simultaneously with audio information;
(2) by receiving the switching events coming from the patients;
(3) by setting the application parameters (connection settings, scrolling time, number of iterations).

The server module is represented by a software application that receives data from the client application by using the well-known TCP/IP protocol.

The client application running GUI works as the schematic flowchart presented in Fig. 2.2. After power-up or a reset the system enters into a standby mode and remains inactive until the patient trigger the switch. When this thing happens, the entire list of keywords and associated images/icons begin to scroll on the LCD display with the configured scrolling rate. If the patient desires to select one keyword from the list he/she has to trigger again the switch, and the application works in one of the following conditions:

- if the selected keyword represents the name of a keywords (sub)category (class), that class is selected and the list continue to scroll by using keywords within the selected class;
- if the selected keyword represents a keyword belonging to a (sub)category, then on the screen appears the expression "I SEND".

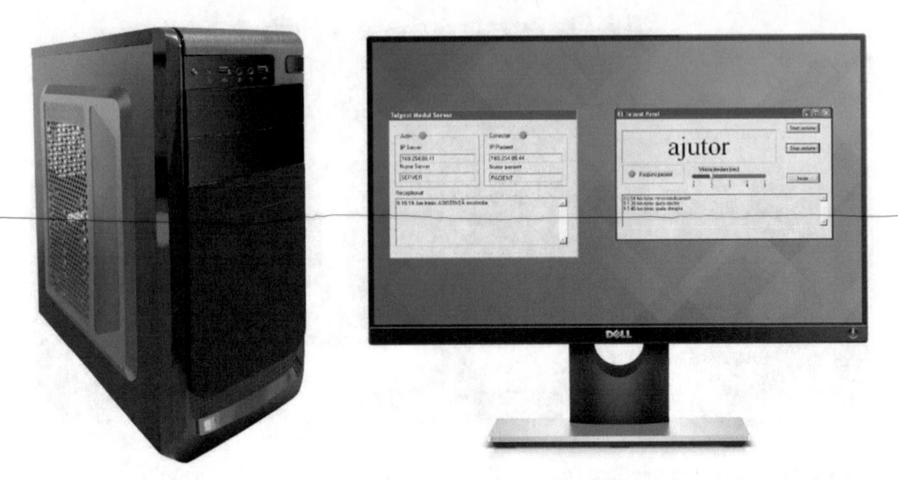

Fig. 2.5 The software running on server—GUI (Romanian word for "help")

The application has a cyclical function as long as the number of iterations for the entire list of keywords, configured at the startup, for each (sub)category, is not outrun.

When the sending of the keyword is confirmed by the patient, by triggering again the switch, the selected keyword and its subcategory are stored into a "list of selected words", together with the co-ordinates of the time. Same information is forwarded to the server application for further processing and storing (Fig. 2.5). Figure 2.5 shows also an example of a screen as produced by the patient application in the server for the word "Help".

The TELPROT system has been tested with good results on 27 neuromotor disabled patients from "Prof. Dr. N. Oblu" Emergency Hospital of Iasi, Romania and the obtained results revealed the necessity to use very simple dialogs between patients and caretakers, by simply answering with YES/NO to the questions instead of using keywords.

The described system has been used a very short period and improved further by increasing the number and complexity of the key phrases, introduce extra switches/software commands, so that the patient should have more than two options, etc.

2.4 Interactive Communication System for Patients with Disabilities—ASISTSYS Project

In the last few years it has been observed an increasing interest for the improvement of communication with people with disabilities for therapeutic and educative purposes. Following this issue, a second variant (improved) of the communication system, called ASISTSYS, has been implemented and tested. The system was based as previous TELPROT system on keyword transmission from the disabled patient and vocal and/or keyword reply from the caretaker.

The developed system allows communication between patient and caretaker in the following situations:

(a) in case of treatment within hospital or care center for recovery through a dispatch;
(b) in case of disabled patient treatment in their home.

The overall architecture of the ASISTSYS includes three main components (Fig. 2.6): patient module, caretaker and dispatcher.

The ASISTSYS system is made up of two or three subsystems, depending on where it is used (care center/hospital or homecare). In the case of a care center the following hardware components are used: a laptop for each disabled patient, a PDA/Mobile phone (Smartphone) for the caretaker, and a computer (usually a PC) for the dispatcher are used (Fig. 2.6). The homecare configuration does not use a dispatcher and includes only the laptop and Smartphone (Fig. 2.7).

On the patient's screen the keywords and associated images are simultaneously displayed and played on the speakers.

In order to allow the patient to select the desired keyword, ASISTSYS system uses the following selection techniques:

(1) The technique of using a mechanical two positions switches controlled by the patients—for patients who can perform certain controlled muscle contractions;
(2) The technique based on the use of electrooculography (EOG) by measuring the corneo-retinal biopotential that exists between the front and the back of the human eyes. The selection event is triggered by blinking the eyes;

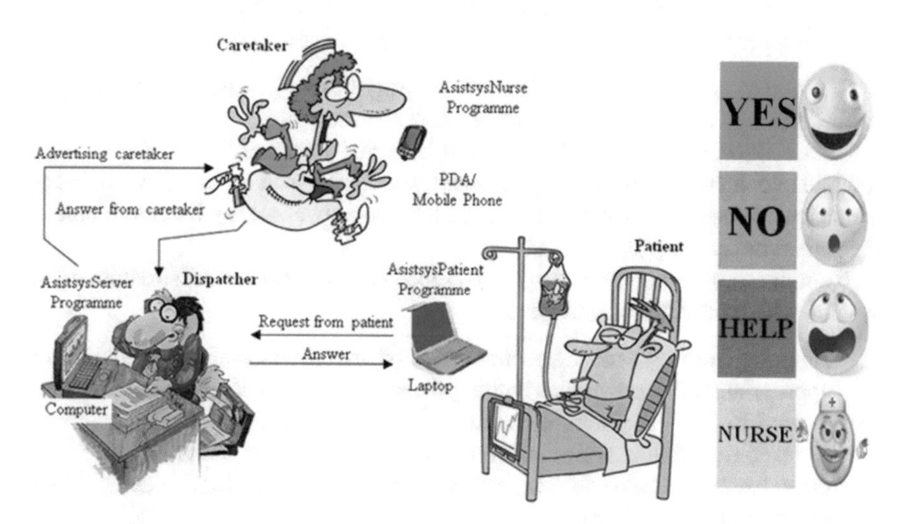

Fig. 2.6 The overall architecture of ASISTSYS medical communication system

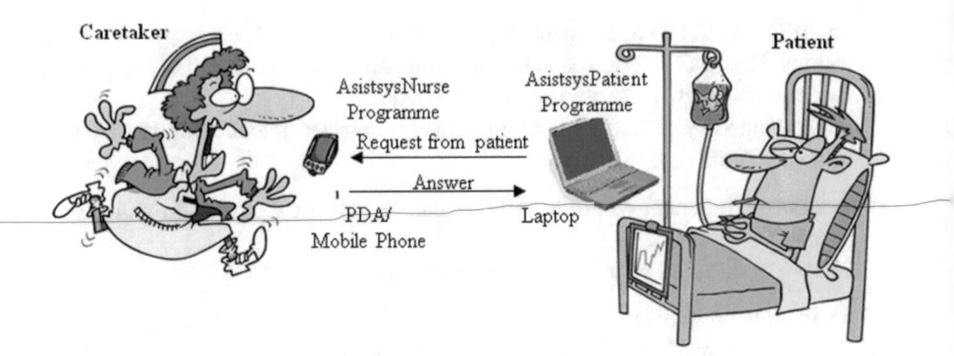

Fig. 2.7 The homecare configuration structure

(3) Image-based technique by using a video camera connected to patient laptop that process the images of the eyes and compute position of the pupils. The selection event is triggered by the eyeballs movements.

The software applications running within ASISTSYS system allows the disabled patients to be assisted and contains the following modules:

(a) AsistsysPatient—a software module used to assist the patient in transmitting the selected keyword to the dispatcher. It runs on the patient's laptop and interacts with he/she by using the screen for displaying the keywords/images and the triggered event (by using one of the previous described technology);
(b) AsistsysServer—a server software module running on the dispatcher PC and used in bidirectional communication with ASISTSYSPatient and ASISTSYS-Nurse modules;
(c) AsistsysNurse—a software application running on Smartphone and used by the caretaker to receive notification and to reply to the patient.

The communications AsistsysPatient—AsistsysServer—AsistsysNurse are performed through the radio via Internet resources: through the WLAN and/or mobile Internet (GSM/GPRS) network. It should be noted that both WLAN and GPRS (mobile telephony) resources are permanently accessible and accessed as needed by the status of one or other network (one of the two networks may be unavailable to the patient or caretaker).

It emphasizes the persistence of the presence of a PC server in the structure of the system, which provides essential facilities:

(1) Automatic resource management of the system: (a) choosing the available and appropriate radio network (b) calling the available caretaker;
(2) Record history of system operation: time and type of requests, time and type of responses, failures or other events. History of dialogs between patient and caretaker is accessible for verification, which is the basis for achieving a high quality of service (QoS).

The principles of operations in ASISTSYS system is represented in Fig. 2.8. By triggering the appropriate switch in the standby mode the patient initiates a request (physiological needs, desires, etc.) assisted by AsistsysPatient module. The request is received and processed by AsistsysServer module and forwarded to the caretaker responsible for the selected patient. The caretaker can respond to the request by pressing a YES/NO button on the AsistsysNurse module user interface. The reply is sent back to the patient through AsistsysServer module and if the answer is affirmative. Otherwise, in case of a negative answer from the caretaker is provided to the AsistsysServer module and another caretaker available in the network will be notified. Similar behavior of the system can be found in the situation when the connection between AsistsysServer and AsistsysNurse modules cannot be established.

Keywords used in ASISTSYS have been grouped in three levels:

(1) First level is the lowest level and contains only two words (YES/NO). It is used in the early stages of disabled patients training;

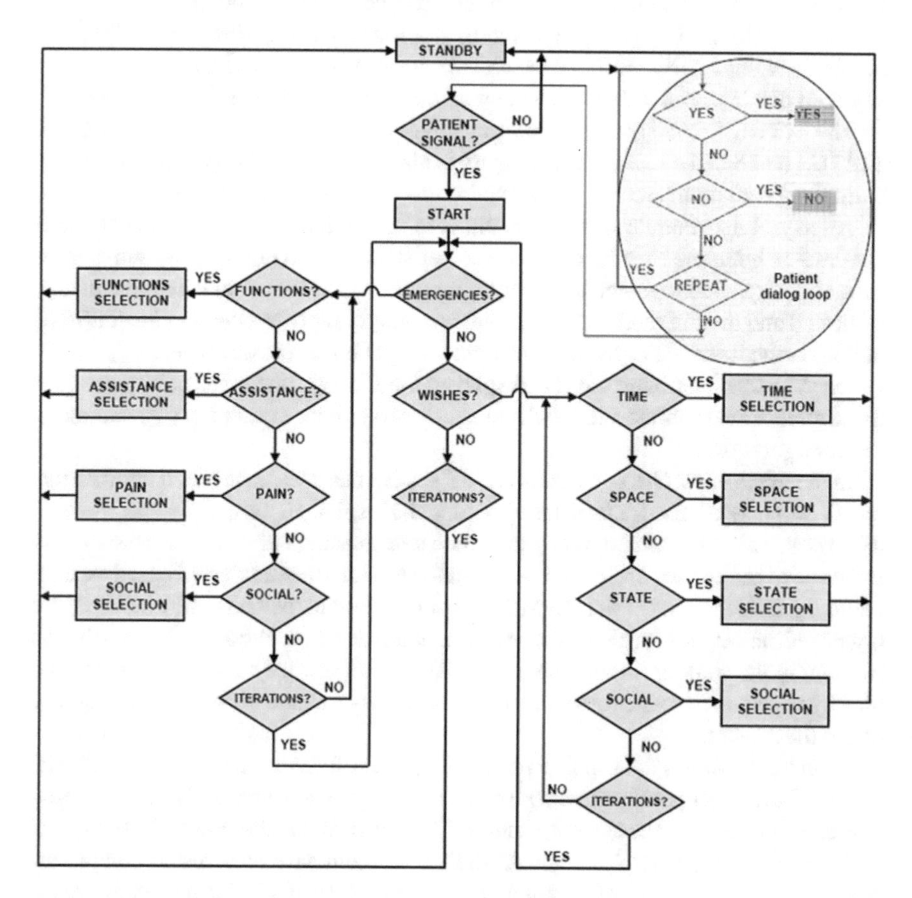

Fig. 2.8 Functional diagram with patient dialog loop (AsistsysPatient)

(2) The second level containing 10–12 keywords of key interest (such as ‘it hurts’, ‘can’t breathe’, …) and used in the second training phase. These words can be edited according to the particular condition and ability of each disabled patient.
(3) The third level containing 50–100 words, grouped on categories, usable by disabled patients with good comprehension skills for complex communications. These words can also be edited according to particular needs.

In Fig. 2.8 it is represented the functional diagram of ASISTSYS system with patient dialog loop running on the laptop screen and included in the AsistsysPatient module. At first event triggered by the patient on the laptop screen are displayed two slides EMERGENCY/WISHES marked in the background by red or green color and a graphical representation of a keyword referring to the category of concepts. The graphical representation (pictogram) has the role of making the choice easier for patients that may have difficulties in reading the keywords. When a keyword from a slide is selected, the patient hears a voice naming the choice: EMERGENCY or REQUEST. If the patient has made a wrong choice, it can be undone.

The EMERGENCY category contains the four subcategories: NECCESITIES, ASSISTANCE, PAIN, SOCIAL. Each subcategory represented by a pictogram, a keyword displayed on the screen and an audio recording of the selected keyword.

The WISHES category contains the following subcategories: TIME, SPACE, STATUS and SOCIAL. Each subcategory is also represented by a pictogram, a keyword displayed on the screen and an audio recording of the selected keyword.

AsistsysNurse module allows a caretaker to be notified by patients in his/her care. On the user interface running on the caretaker smartphone (Fig. 2.9) the name of the patient is displayed together with information related to the room where the patient is located and their request. The caretaker can simply reply by pressing the YES/NN button on the interface or by also writing to the patient a text message.

On the same user interface it is displayed the status of the connectivity between the caretaker smartphone and the dispatcher’s server/patient’s laptop (depending on the configuration used).

In the first testing phase, the ASISTSYS system has been tested with good results on 31 subjects at the Rehabilitation Clinical Hospital in Iasi, Romania. Several databases for keywords and pictograms with a decreasing number of necessary keywords have been used. The patients used in the test have been selected based on their neurological deficiency. For better-trained or cooperating patients, the databases are organized on several levels of hierarchy, making it easier choosing keywords and shortening the waiting time of the patient All the keywords and icons are accompanied by appropriate sound messages that make it easy for patients with high disability to use the system.

Figure 2.10 shows the training phase of a patient for the use of ASISTSYS. The patients have been positively receptive to such a system, some of them even being interested on the date the system would be permanently integrated into hospital use.

In the second testing phase, the ASISTSYS system have been tested “Bagdasar-Arseni” Emergency Clinical Hospital in Bucharest, Romania, being positively appreciated by patients, clinicians, medical doctors and nurses. Figures 2.11, 2.12, 2.13 and 2.14 presents the tests performed with the system used by the trained disabled patients.

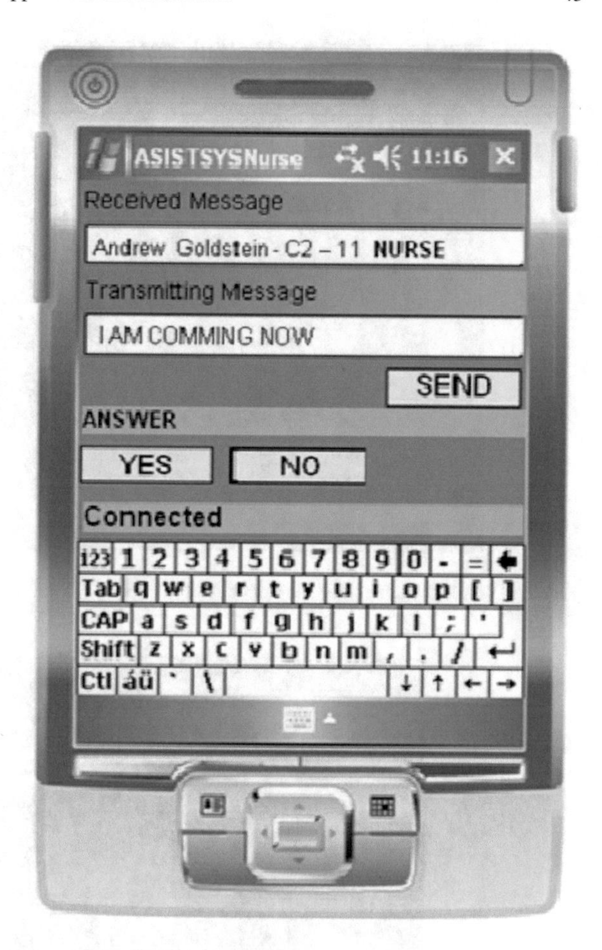

Fig. 2.9 AsistsysNurse screen

2.5 A New Medical Device for Communication with Neuromotor Disabled Patients

The device has been developed as a new module for assistive systems, being useful for those patients that are able to use a virtual keyboard and desire to play a video game. The medical device contains a custom designed electronic parts connected to a microcontroller development board for the electrooculography signals acquisition and processing and a tablet PC running a software application in form of a visual keyboard. The device is designed to be used by the neuromotor disabled patients that are able to move the eyeballs. The application running on microcontroller detects the movements of the patients eyeballs in four directions and selects a key from a virtual keyboard by using dwell time selection method. The proposed medical device has been developed and tested on 10 simulated patients in order to test its performances in terms of key selection accuracy (Fig. 2.15).

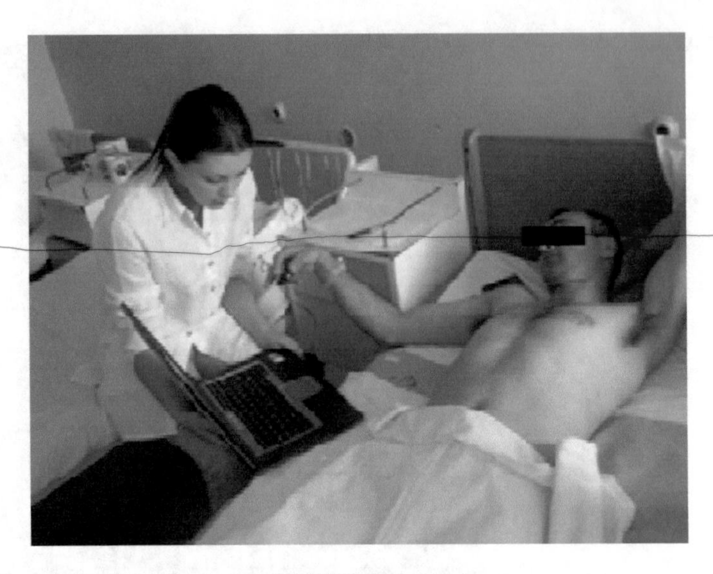

Fig. 2.10 Patient training for the use of ASISTSYS

Fig. 2.11 The AsistsysPatient module

(a) the data acquisition interface, a custom developed board, containing all necessary sensors, cables amplifiers and filters connected together and used for acquiring the electrooculography (EOG) signals.
(b) a microcontroller board (Arduino Leonardo) responsible for digitizing the horizontal and vertical EOG signals, processing in order to detects movements and transmitting the commands (move left, move right, move up and move down) to PC.
(c) the PC running a virtual on screen keyboard (part of Windows OS) used to display successive words written by the disabled patient.

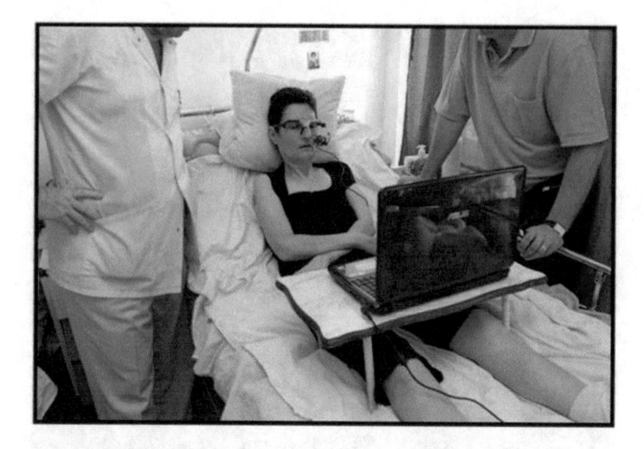

Fig. 2.12 Patient testing the eye—tracking device

Fig. 2.13 Testing AsistsysNurse module

The microcontroller board use the A/D channels to sample the amplified EOG signal with a sampling frequency of 500 Hz and 10 bits/sample. The software running on microcontroller process the EOG signals and detects the eye movements in the four directions: left, right, up, and down according to the flowchart presented in Fig. 2.16. The detection method uses two adaptive thresholds for each channel (horizontal and vertical) by computed by information provided by the output of the low pass filter.

The prototype of medical device for communication with disabled patients based on EOG signal processing and microcontroller development board (Fig. 2.17) is presented in Fig. 2.18.

The character selection method used in communication between patient and clinicians is performed by selecting characters from an alphabet list and building

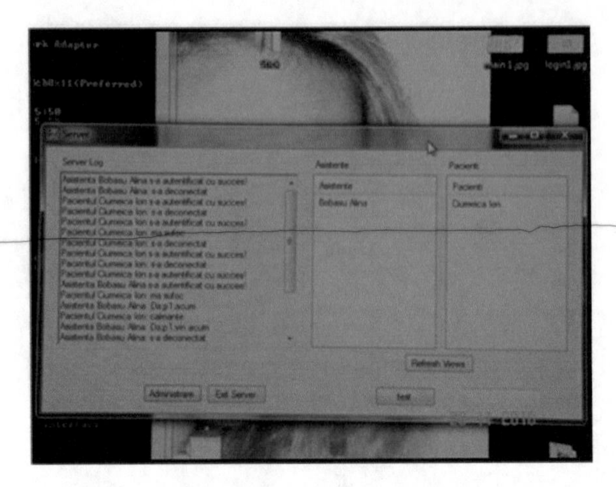

Fig. 2.14 The AsistsysServer module

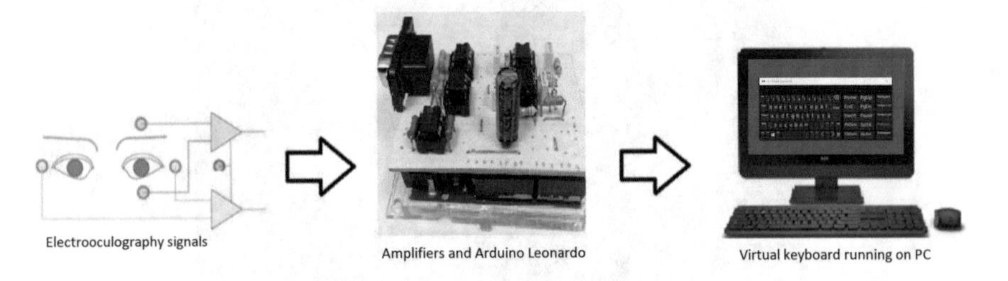

Fig. 2.15 Overall architecture and integration of the proposed medical device

free sentences using them. This method results in complex messages, but, in order to obtain good results, the method requires patient skills for working properly with this device.

The choice of using dwell time selection method was a tradeoff between speed and the rate of selection: the shorter the dwell time, the faster keys from keyboard can be selected, but the higher the rate of unintentional selection.

The classification results obtained are compared in terms of the classification accuracy related to the time needed for each selection of the desired key. The experiments have been performed on 10 simulated patients, asked to select a specific key from keyboard in a period of time less than 3 s, including dwell time, after a short training period of time. The averaged accuracy obtained for all patients was 90%, value that makes the device suitable to be used in practice.

The practical approach of using the proposed medical device as a controller for video games is represented in Fig. 2.19.

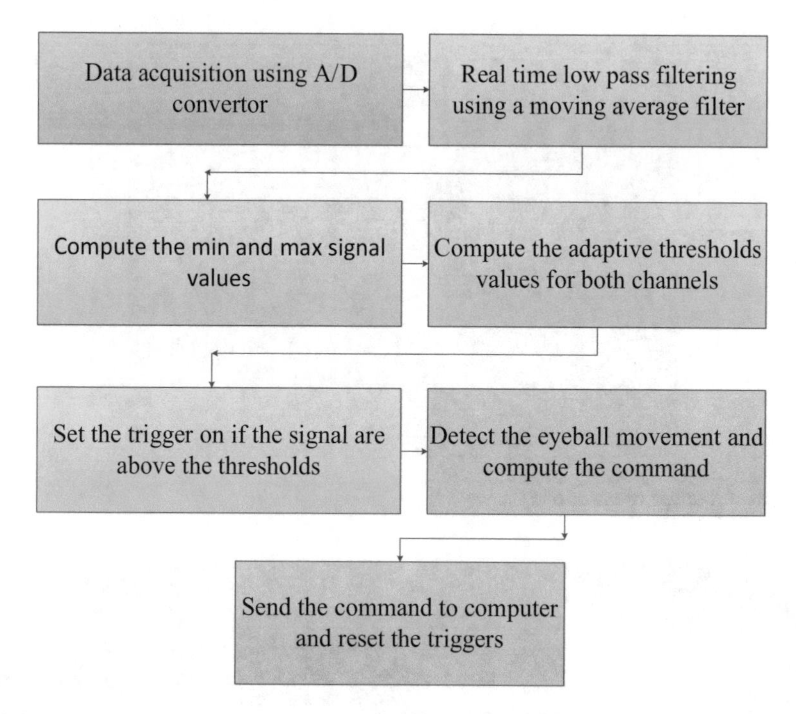

Fig. 2.16 The flowchart of the software running on microcontroller

Fig. 2.17 Microcontroller development board

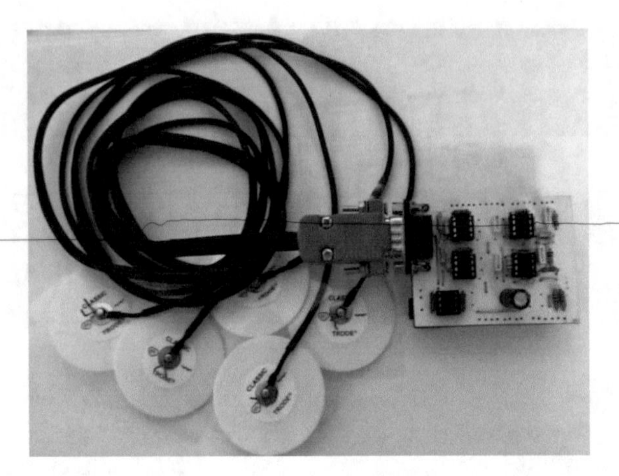

Fig. 2.18 Prototype of the device

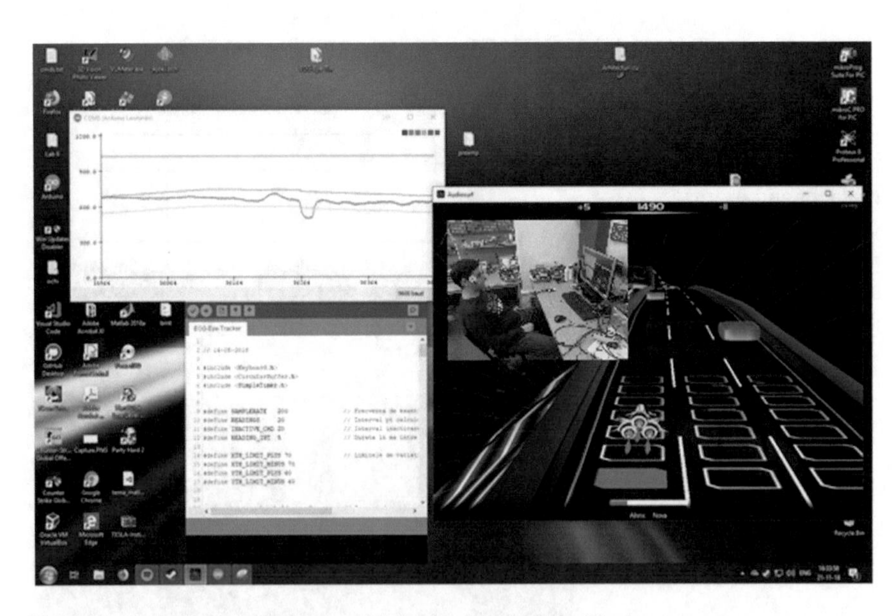

Fig. 2.19 Controlling a video game with the developed medical device

2.6 Integrated System for Assistance in Communicating with and Telemonitoring Severe Neuromotor Disabled People—SIACT

The final variant of the project proposed to carry out the necessary research, to design and build a complex electronic system for assisting in the communication with and remote monitoring of the physiological parameters of patients with major neuromotor disabilities.

The solution proposed by this project consisted in the design of a complex system capable of ensuring both communication and monitoring of essential physiological parameters needed for treatment and caretaking. It is composed of two functional components: one for communication and one for monitoring, and they will share the same hardware platform.

The hardware platform includes: (1) a patient subsystem made of a laptop/Smartphone/tablet for communication and a number of wireless sensors for remote monitoring; (2) a workstation acting as a server/dispatch; (3) a caretaker subsystem (Smartphone/tablet PC) than ensures access to the patient subsystem via the server. The communication function is bidirectional (patient ↔ caretaker through server). The communication uses the keywords technology: the system ensures successive visualization/audio playback of keywords/ideogram/alphanumeric characters and the patient giving to the disabled patient the possibility of keyword selection by using different methods according to their physical condition. The selected keyword is then sent to the caretaker. The detection technique can be performed either by using a switch attached to the patient's forearm/finger/leg or by detecting the patient's eye movement using EOG signals or video analysis. The system will allow patients to use computers to communicate with other people, but also to read newspapers, books and to use the Internet. The caretaker's response will be sent to the patient in the form of sound and/or visual as text/ideogram (Fig. 2.20).

The remote monitoring of physiological parameters is done with the purpose of efficient and prompt assessment of the patient's state, allowing rapid intervention in case of necessity. The system will have the ability of sending alarms to the caretaker when limit values are exceeded and will also allow continuous observation of these parameters. All monitored parameters, alarms and caretaker interventions will be recorded in the server memory for later analysis.

This system presents a series of clear advantages: (1) from the patient's point of view, psychic comfort is increased by ensuring communication with the surrounding

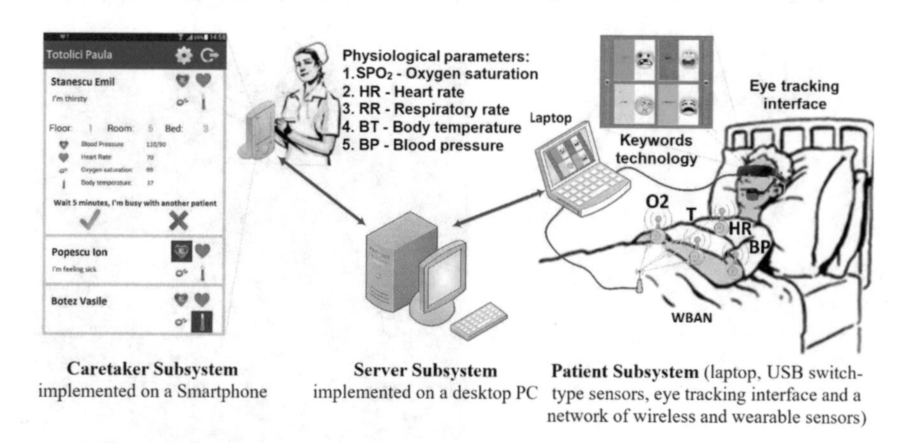

Fig. 2.20 The architecture of the integrated system for assistance in communicating with and remote monitoring disabled people

environment; (2) medical investigation is facilitated by the possibility of communication and also due to biomedical parameters monitoring; (3) prompt interventions are possible in the case when vital parameters become abnormal; (4) increased accountability of medical personnel given by the logging of caretaker interventions; (5) lower caretaking costs due to a smaller number of caretakers being able to assist a larger number of patients.

The end product is intended for hospitals, care and treatment centers, nursing homes and also for patient residences.

The proposed assistive system uses two types of a real-time eye tracking interfaces to communicate with severely neuromotor disabled patients by using keywords technology:

- head-mounted device (Fig. 2.21), which measures the angular position of the eye from the head;
- remote device (Fig. 2.22), which measures the position of the eye to the surrounding environment, implemented with commercially available interface developed by TOBII; the IR sensors are placed at the base of the screen.

The head-mounted eye tracking interface, illustrated in Fig. 2.21 consists of an infrared video camera mounted on a frame glasses right underneath the eye, connected to a patient subsystem (laptop), for eye pupil image acquisition and processing. The subjects who tested the system were asked to place their head in a chin rest and look at the user screen placed approximately 60 cm away. This communication subsystem is used by patients with severe neuromotor disabilities who cannot communicate with the outside world through classical methods: speaking, writing, hand gestures (signs).

The remote device presented in Fig. 2.22 is used by patients who, due to their affection, do not support a helmet on their head.

The communication function is bidirectional (patient $\leftrightarrow$ server $\leftrightarrow$ caretaker). The patient $\leftrightarrow$ caretaker communication relies on key-words tech-

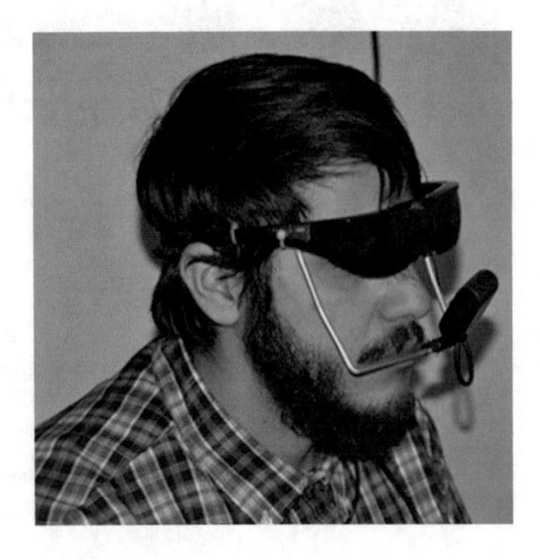

Fig. 2.21 Head-mounted eye tracking interface

Fig. 2.22 Remote eye tracking interface

nology: the system ensures successive visualization/audio playback of keywords/ideogram/alphanumeric characters (by using a virtual keyboard), and then the patient has the possibility of choosing the desired keyword.

The remote monitoring module is based on wireless and wearable sensor network, where each node contains a sensor for acquiring the physiological parameters of the disabled patients, according to their needs (Fig. 2.23). Thus, each device in the network ("Sensor Node"—SN) is wirelessly connected to the Central Node, which is USB connected with the patient's calculation unit and via this—to the Server. The structure of this function is based on internet of things (IoT) concept, by using interconnected sensors in distributed network. The measurements performed by the sensors are efficiently transmitted to the server, that is also responsible for the management of the entire sensors network. After local processing of the measured parameter, the server transmits the result to the mobile caretaker devices.

The remote monitoring subsystem contains a wireless body area network (WBAN) of medical devices attached to the patient's body for the acquisition of the following physiological parameters: blood oxygen saturation (SpO_2), heart rate (HR), respiratory rate (RR) and body temperature (BT). The number and type of the monitored parameters can be adapted to the medical needs of patients and will be selected based on medical recommendation.

The devices used to remote monitoring of the disabled patients (Fig. 2.23) contain commercial electronic modules used to acquire physiological parameters and having as a feature the low energy consumption. These devices are connected to wireless data transmission/reception modules (eZ430-RF2500) in order to transmit the measurements.

The computer of the patient runs a background software application that retrieves the data collected by medical devices and sends them to the server for parameters processing and is then transmitted wirelessly to the caretaker device where they are displayed in real time.

If it detects the exceeding of certain limits considered dangerous, the Server, by automatic data analysis, alerts the caretaker and his interventions will be recorded in the server memory for later analysis. The values of the monitored data can be

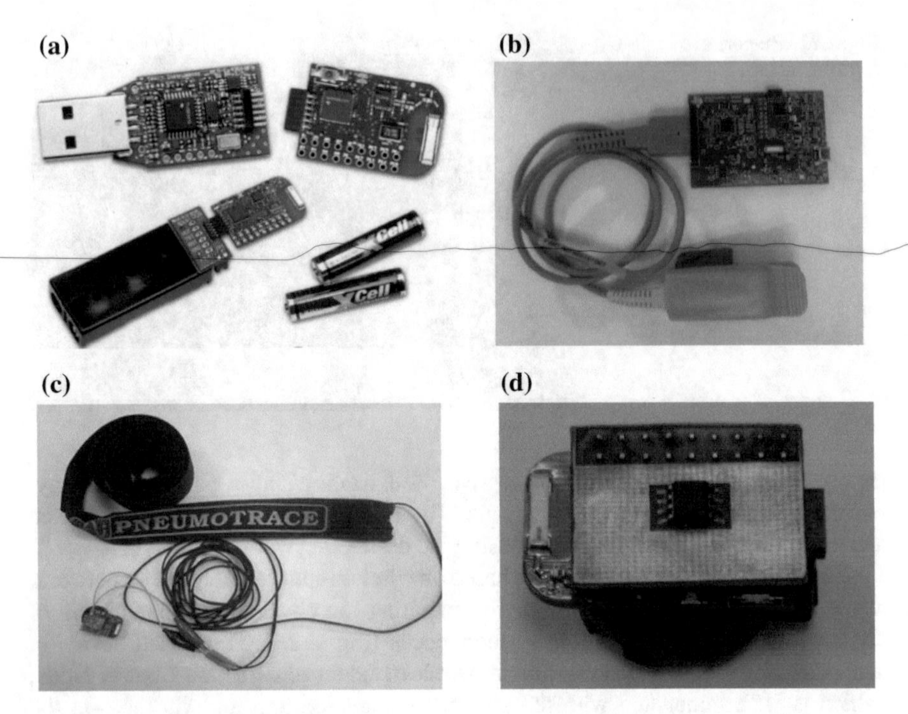

Fig. 2.23 eZ430-RF2500 wireless development kit (microcontroller and wireless transceiver) (**a**); device for telemonitoring heart rate and blood-oxygen saturation (**b**); respiratory rate (**c**) and body temperature (**d**)

processed at any time to see the evolution of the time of the physiological parameters of the patients, a very useful feature for establishing the optimal treatment for the patient.

The three components of the proposed support system (patient, server, caretaker) are connected together via the internet. The operating modes of the system components are controlled by the software.

The Patient WEB application (Fig. 2.24) has been designed to be robust, flexible and easily adaptable to the necessities of each disabled patient, by using switch-based or eye tracking-based keywords selection methods.

The software component used to detect the movements of the disabled patient's eyes is based on the implementation of a custom developed algorithm for eye tracking and selection of ideograms/keywords the in real time, including video camera calibration, pupil center detection and mapping of the coordinates within the image provided.

To improve the pupil position detection, the well-known dark pupil technique has been used. Thus, by using infrared illumination, the pupil is very well emphasized to iris and sclerotic, appearing in the image as a very well-defined black area. On the other hand, by using this type of illumination (infrared), uniform and consistent illumination of the eyes is obtained without any discomfort from the subject/patient.

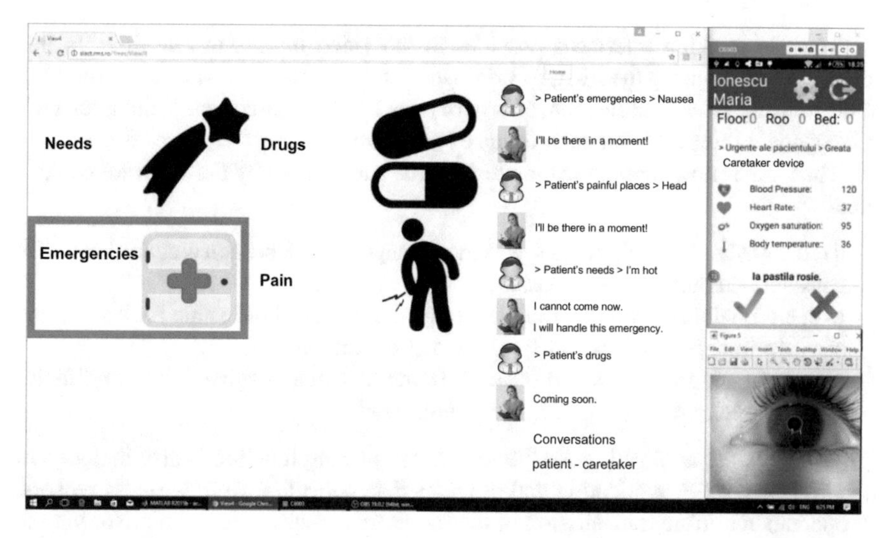

Fig. 2.24 WEB application for communication using switch-based or eye tracking-based patient needs detection (patient device screen—Laptop, caretaker device screen—Smartphone, raw eye image provided by the IR video camera)

Regardless of the method used this software component includes the following:

1. Acquisition of real-time eye images using an infrared camera;
2. Calibration of the system:
   - nine target points for determining the mapping function coefficients;
   - focused on pupils, in order to determine the optimal parameters of the algorithms used for image binarization;
3. Real-time pupil center detection (for each frame provided by the infrared video camera) using different pupil detection algorithms (ADP);
4. Mapping between the detected pupil center in the eye image and the corresponding position of the cursor on the user screen;
5. Selecting ideograms and/or keywords by simulating a "mouse click" through extended eye point at a fixed point.

Generally, the pupil detection algorithms (ADP) include the following steps:

1. Acquisition of real-time eye image using an infrared camera;
2. Filtering the image of the eye;
3. Binarization of the image of the eye by segmentation of fixed/global/local adaptive threshold image;
4. Removing/diminishing corneal reflection;
5. Extracting contour pupils detected;
6. Detection of pupil center—using a certain technique imposed by ADP;
7. Optimize the algorithm by stabilizing the cursor on the user's screen (real-time filtering of ADP signals, removal of parasitic pulses and snap-to-point technique.

The eye tracking interfaces provides to the video images in real-time. In this case, because usually the image's brightness and contrast may vary from frame to frame during system utilisation, it is necessary to use binarization techniques with independent thresholds for each frame of video images.

Thus, the following binarization techniques can be used by the software component:

- fixed threshold for all frames of video images—in laboratory conditions with constant and uniform illumination;
- with a global threshold that modifies (adjusts) the threshold value for each frame of video images—for normal system usage conditions;
- with local adaptive threshold for each frame of video images—using in difficult operating conditions (non-uniform illumination).

The software application for the remote monitoring function is also included in the Patient's WEB application and provides data collection from wireless sensors. It operates real-time transmission of the monitored values to the supervisor via the server. The remote monitoring feature includes a graphical interface running on the patient computer and it used for real-time numeric and real-time display of monitored physiological parameters, alerts resulting from their processing, and status of each node in the network (voltage at battery terminals powered). The communication protocols used for transmitting monitored data on the server are also included.

In the case of switching-based communication, Patient's WEB application performs ideogram scrolling on the user's screen by their cyclical highlighting (Fig. 2.24). The patient can select an ideogram that acts on the switching sensor only during the time it is highlighted. This time interval is determined by the patient's experience in using the system. The ideogram/keyword database is organized hierarchically. Each level in the database contains a set of ideograms/keywords belonging to a class of objects defined at the previous hierarchical level (ascending). When the patient selects an ideogram using the switch, his request is transmitted to the caregiver via the server.

In order to alert the caretaker when the values of the remote monitored physiological parameters of the patient are exceeded above normal values, a simple alert detection algorithm has been developed.

The physiological conditions that may cause alerts are: low SpO_2 if SpO_2 is less than 93%, heart bradycardia if HR below 40 bpm, heart tachycardia if HR more than 150 bpm, HR arrhythmia if ΔHR/HR over last 5 min is greater than 20%, HR variability if max HR variability is higher than 10%/the last 4 readings, hypothermia if BT drops below 35 °C, hyperthermia if BT is elevated beyond 38 °C, low respiratory rhythm if RR falls below 5 rpm, low battery voltage if VBAT less than 1.9 V, low value for RSSI if measured RSSI is less than 30%.

To assess the functionality of the proposed assistive system, the 30 patients who participated to system testing at the "Dr. C. I. Parhon" University Hospital of Iaşi, Romania from Clinic of "Geriatrics and Gerontology" answered to a questionnaire which focused on the necessity, usefulness, operating modes, flexibility and adaptability of the system to the patient's needs. Patient response was assessed on a scale

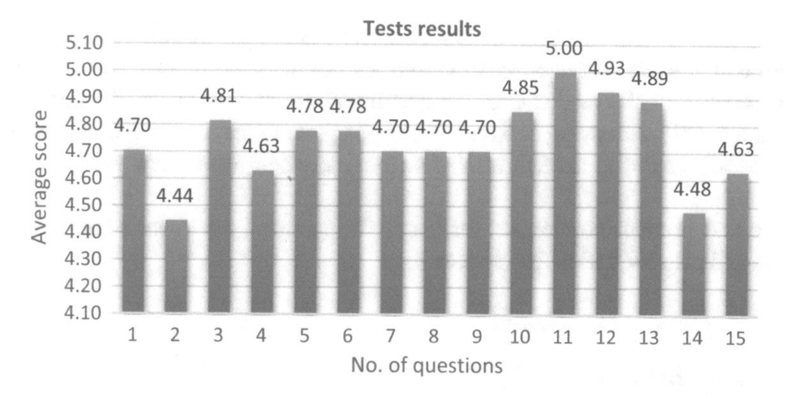

Fig. 2.25 Average score of patients' assessment who participated to system testing at the "Dr. C. I. Parhon" University Hospital of Iaşi, Romania

between 0 (lowest level of appreciation) and 5 (highest level of appreciation), and the value corresponding to the patient's best estimate was scored in the response table (Fig. 2.25).

The proposed system presents a series of clear advantages compared to other similar ones: (1) the innovative principles of two-way communication between severely disabled people and caretakers/medical staff by gaze detection techniques based on video oculography; (2) the possibility of medical investigation of patients with severe neuromotor disabilities who cannot communicate with other people by means of classical communication (spoken, written or by signs) by using the communication technique through keywords/ideograms organized on several hierarchical levels; (3) the possibility of implementing advanced communication by alphanumeric characters, which allows the severely neuromotor disabled patients to use the Internet and e-mail; (4) the possibility of adapting both software and hardware structure on the way, according to the patient's needs, their evolution and medical recommendation; (5) the permanent monitoring of several physiological parameters, their analysis and the alarming of caretakers in emergency situations—all these done using the same resources as those for communication; (6) increased accountability of medical personnel given by the logging of caretaker interventions; (7) lower caretaking costs due to a smaller number of caretakers being able to assist a larger number of patients.

For obtaining an optimal solution for the prototype, many attempts and tests (both in laboratory and with patients) have been performed. The final tests of our system prototype have been performed at "Dr. C. I. Parhon" University Hospital of Iaşi, Romania, with the patients from Clinic of "Geriatrics and Gerontology".

The test results confirmed the performance of the system, both for communication with neuromotor disabled patients and for telemonitoring their physiological parameters. Both medical staff and patients involved in testing have positively appreciated its functions, easy to use and adaptability of the system to the needs of patients, highlighting its utility within a modern medical system.

The main benefits brought by our assistive system for disabled people, for the health and welfare of the population and also to the state budget are: an increase in the quality of the medical act, a rapid insertion into society of the neuromotor disabled people, the saving of lives in emergency medicine, the decrease in expenses for patients, the decrease of number of medical services given in ambulatory medical care.

The end product is intended for hospitals, care and treatment centers, nursing homes and also for patient residences.

2.7 Conclusions

In conclusion it may state that the essential novelty element was the idea that people with severe neurological disabilities could communicate both ways, to and from the surrounding world. To our best knowledge, there is no system similar to ours. The project is quite complex and interdisciplinary, and its implementation needs a good harmonization of requirements from very different fields: electronics and telecommunications engineering and technology, computer science, neurology, neuro-motor recovery, linguistics, psychology and logopedy. To choose the best solutions, many tests and difficult work with patients are necessary. The testing methods and methodologies that will be used have to be first elaborated and trialed, using existing equipment (PCs, radio transmitters and receivers), before constructing system models. Also, a number of experimental blocks should be constructed and tested before choosing the final solution for the prototype.

The system substantially improves the general living conditions of the disabled patients, as:

1. it allows automatic communication between patient and caretaker, including the transmission of several complex messages;
2. it removes the need of the constant presence of the attendant near the patient;
3. it eases the patient discomfort and the work burden of the caretaker, who can "find out" what the patient needs;
4. the same caretaker can attend to several patients, without any loss in quality care;
5. the associated care costs are diminished, with no reduction of the care quality.

References

1. Cook, A.M., Hussey, S.M.: Assistive Technologies: Principles and Practice, Pb. Mosby-Year Book (2001). ISBN 0323006434
2. Bryant, D.P., Bryant, B.R.: Assistive Technology for People with Disabilities, Pb. Allyn & Bacon (2002) ISBN 020532715X
3. European Commission: A Strategy for Smart, Sustainable and Inclusive Growth. COM (2010). www.ec.europa.eu/eu2020

4. Bozomitu, R.G.: New methods of detecting voluntary blinking used to communicate with disabled people. Adv. Electr. Comput. Eng. **12**(4), 47–52 (2012). https://doi.org/10.4316/AECE.2012.04007
5. Cehan, V., Cehan, A.-D., Bozomitu, R.G.: A new technology of communication with people with major neuro-locomotor disability. In: Proceedings of the 2nd Electronics System-Integration Technology Conference (ESTC-2008), Greenwich, London, UK, 1st–4th Sept 2008, pp. 791–795. IEEE Catalog Number: CFP08TEM-PRT, ISBN 978-1-4244-2813-7
6. Lupu, R.G., Ungureanu, F., Bozomitu, R.G.: Mobile embedded system for human computer communication in assistive technology. In: Proceedings of the Intelligent Computer Communication and Processing (ICCP 2012), IEEE International Conference, Sinaia, Romania, Aug 30–Sept 1 2012, pp. 209–212. ISBN 978-1-4673-2951-4
7. Bozomitu, R.G., Barabaşa, C., Cehan, V., Lupu, R.G.: The hardware component of the technology used to communicate with people with major neuro-locomotor disability using ocular electromyogram. In: Proceedings of the 17th International Symposium for Design and Technology of Electronic Packages, SIITME 2011, Timişoara, Romania, 20–23 Oct 2011, pp. 193–196. ISBN 978-1-4577-1275-3
8. Cehan, D.A., Lupu, R.G,. Cehan, V., Bozomitu, R.G.: Key-word data base used in communication system with disabled people. In: Proceedings of the 17th International Symposium for Design and Technology of Electronic Packages, SIITME 2011, Timişoara, Romania, 20–23 Oct 2011, pp. 365–368. ISBN 978-1-4577-1275-3
9. Lupu, R.G,. Bozomitu, R.G., Ungureanu, F., Cehan, V.: Eye tracking based communication system for patient with major neuro-locomotor disabilities. In: Proceedings of the 15th International Conference System Theory, Control, and Computing (ICSTCC 2011), Sinaia, Romania, 14–16 Oct 2011, pp. 318–322. ISBN 978-973-621-322-9
10. Cehan, V., Cehan, D.A., Bozomitu, R.G., Lupu, R.: A new system for communication with disabled people. In: Proceedings of the Digital Cities Symposium, ERA-6 Conference, Athens, Greece, 23–24 Sept 2011
11. Lupu, R.G., Bozomitu, R.G., Cehan, V., Cehan, D.A.: A New computer-based technology for communicating with people with major neuro-locomotor disability using ocular electromyogram. In: Proceedings of the 34th International Spring Seminar on Electronics Technology, ISSE 2011, High Tatras, Slovakia, 11–15 May 2011, pp. 442–446. ISBN 978-1-4577-2111-3
12. Bozomitu, R.G., Cehan, V., Barabaşa, C.: A new VLSI implementation of a CMOS frequency synthesizer for SRD applications. In: Proceedings of the 16th International Symposium for Design and Technology of Electronic Packages, SIITME 2010, Piteşti, Romania, 23–26 Sept 2010, pp. 181–186. ISBN 978-1-4244-8123-1
13. Cehan, V., Bozomitu, R.G., Barabaşa, C.: Interactive communication system with patients with disabilities—the software component. In: Proceedings of the 16th International Symposium for Design and Technology of Electronic Packages, SIITME 2010, 23–26 Sept 2010, Piteşti, Romania, pp. 323–326. ISBN 978-1-4244-8123-1
14. Bozomitu, R.G., Cehan, V., Barabaşa, C.: A VLSI implementation of a 3 Gb/s LVDS transceiver in CMOS technology. In: Proceedings of the 15th International Symposium for Design and Technology of Electronic Packages, SIITME 2009, Gyula, Hungary, 17–20 Sept 2009, pp. 69–74. ISBN 978-1-4244-5132-6
15. Barabaşa, C., Bozomitu, R.G., Cehan, V.: Application of short range radio devices in communication systems for patients with severe neuro-locomotor disabilities. In: Proceedings of the 15th International Symposium for Design and Technology of Electronic Packages, SIITME 2009, Gyula, Hungary, 17–20 Sept 2009, pp. 251–255. ISBN 978-1-4244-5132-6
16. Cehan, A., Radinschi, I., Cehan, V., Bozomitu, R.G.: Interactive system for assisting the educational process of students with special requirements. In: Vol. 3 of Research, Reflections and Innovations in Integrating ICT in Education, pp. 1358–1360. Zurbaran 1, 2ª Planta, Oficina 1, 06002 Budajoz, Spain, BA-224-2009 (The V International Conference on Multimedia and Information & Communication Technologies in Education—m-ICTE2009, Lisbon, Portugal). ISBN of Collection: 978-84-692-1788-7, ISBN vol. 3: 978-84-692-1791-7

17. Barabaşa, C., Cehan, V., Bozomitu, R.G.: Application of short range radio devices in biological signal monitoring, in advancements of medical bioengineering and informatics. In: Proceedings of the Second Edition of the International Conference E-health and Bioengineering—EHB 2009, Constanţa, 17–18 Sept 2009, pp. 128–131. "Gr. T. Popa" University of Medicine and Pharmacy Publishing House, Iaşi, 2009ISSN 2066-7590
18. Lupu, R.G., Cehan, V., Cehan, A.: Computer-based Communication system for people with neuro-muscular disabilities. In: Proceedings of the International Carpathian Control Conference ICCC'2009, Zakopane, Poland, 24–27 May 2009, pp. 203–206. ISBN 8389772-51-5
19. Bozomitu, R.G.: Radio Transmitters and Receivers. Ed. Fundaţiei Academice AXIS, Iaşi, Romania (2010). ISBN 978-973-7742-86-5
20. Lupu, R.G.: Medical Telemonitoring and Assistive Technology. Ed. Politehnium, Iaşi, Romania (2013)
21. Cehan, V: The Basics of Radio Transmitters. Ed. MatrixRom, Bucureşti, Romania (1997). ISBN 973-9254-39-X
22. ASISTSYS: Integrated System of Assistance for Patients with Severe Neuromotor Affections. National Multiannual Grant, PNCDI2, Partnerships, within the Program Joint Applied Research Projects, Funded by the Romanian National Authority for Scientific Research (MEN - UEFISCDI), Contract No. 12-122/01.10.2008, 2011 (2008)
23. TELPROT: Communication system with people with major neuromotor disability. TUIASI, CEEX Contract No. 69 CEEX–II 03/28.07.2006/15213/4.08.2006, 2008 (2006)
24. Bhatnagar, S.C,. Silverman, F.: Communicating with nonverbal patients in India: inexpensive augmentative communication devices. http://www.dinf.ne.jp/doc/english/asia/resource/apdrj/z13jo0400/z13jo0405.html
25. New Grant to Fund Research to Aid People with Communication Disabilities, Durham NC, USA, "Duke University Medical Center". http://medschool.duke.edu
26. EnableMart: Producer of communication equipment for the disabled. http://www.enablemart.com
27. Speechview: Producer of software and video systems for the disabled. http://www.speechview.com
28. The Institute on Disabilities at Temple University of Philadelphia. http://disabilities.temple.edu
29. Site of the National Institute of Disability Rehabilitation Research of the US. www.abledata.com
30. TELEMON: Integrated Real Time Monitoring System for Patients and Older Persons. Within the program Joint Applied Research Projects, Funded by the Romanian National Authority for Scientific Research (MEN—UEFISCDI) Contract No. 11-067/18.09.2007, 2010 (2007)
31. Castanié, F., Mailhes, C. et al.: The U-R-safe project: a multidisciplinary approach for a fully nomad care for patients. In: Invited Paper for SETIT 2003, Tunisia, 17–21 Mar 2003
32. Clifford, G.D., Azuaje, F., McSharry, P.E. (eds.): Advanced Methods and Tools for ECG Analysis. Artech House Publishing, Boston/London (2006). ISBN 1-58053-966-1
33. Kropp, A.: Wireless communication for medical applications: the HEARTS experience. J. Telecommun. Inf. Technol. (JTIT) **4**, 40–41 (2005)
34. Malan, D., Thaddeus, R.F. et al.: CodeBlue: an ad hoc sensor network infrastructure for emergency medical care. In: Proceedings of the MobiSys 2004 Workshop on Applications of Mobile Embedded Systems (2004)
35. Rubel, P., Fayn, J., et al.: Toward personal eHealth in cardiology. Results from the EPI-MEDICS telemedicine project. J. Electrocardiol. **38**(4), 100–106 (2005)
36. Varshney, U.: Pervasive Healthcare Computing. Springer (2009). ISBN 978-1-4419-0214-6
37. Yang, X., Hui, C. (eds.): Mobile Telemedicine: A Computing and Networking Perspective. CRC Press (2008). ISBN 978-1-4200-6046-1
38. Bronzino, J.D.: Biomedical Engineering and Instrumentation. PWS Engineering, Boston, MA (1986)
39. Bronzino, J.D.: The Biomedical Engineering Handbook. CRC and IEEE, Boca Raton, FL (1995)

40. Dawant, B.M., Norris, P.R.: Knowledge-based systems for intelligent patient monitoring and the management in critical care environments. In: Bronzino, J.D. (ed.) Biomedical Engineering Handbook, 2nd edn. CRC Press LLC (2000). ISBN 0-8493-0461-X
41. Scherer, M.J.: Living in the State of Stuck: How Technology Impacts the Lives of People with Disabilities, 4th edn. Brookline Books, Cambridge, MA (2005)
42. Rotariu, C.: Systems for Remote Monitoring of Vital Parameters. Ed. Gr. T Popa UMF, Iaşi (2009). ISBN 978-606-544-011-1
43. Rotariu, C., Costin, H., et al.: E-health system for medical telesurveillance of chronic patients. Int. J. Comput. Commun. Control **5**(5), 900–909 (2010)
44. Rotariu, C., Pasarica, A.l., et al.: Telemedicine system for remote blood pressure and heart rate monitoring. In: Proceedings of the 3rd International Conference on E-Health and Bioengineering—EHB 2011, 24–26 Nov 2011, pp. 127–130
45. Rotariu, C., Manta, V.: Wireless system for remote monitoring of oxygen saturation and heart rate. In: Proceedings of the Federated Conference on Computer Science and Information Systems, FedCSIS 2012, Wrocław, Poland, 9–12 Sept 2012, pp. 215–218
46. Rotariu, C., Manta, V., Ciobotariu, R.: Integrated system based on wireless sensors network for cardiac arrhythmia monitoring. Adv. Electr. Comput. Eng. **13**(1), 95–100 (2013)
47. CARDIONET: Integrated system for continuous surveillance in the e-health intelligent network of patients with cardiac diseases (2009). http://www.cardionet.utcluj.ro/Raport_tehnic_et3.pdf
48. TELEASIS: Complex system, on NGN support—next generation networking—for home senior teleassistance (2009). http://www.teleasis.ro
49. Costin, H., Rotariu, C., Dionisie, B., et al.: Telemonitoring system for complex telemedicine services. In: Proceedings of International Conference on Computers, Communications & Control, ICCCC 2006, Baile Felix Spa, Oradea, 1–3 June 2006, pp. 150–155
50. Puscoci, S., Costin, H., Rotariu, C., et al.: TELMES—regional medical telecentres. In: Proceedings of XVII International Conference on Computer and Information Science and Engineering, ENFORMATIKA 2006, Cairo, Egypt, Dec 2006, pp. 243–246. ISSN 1305-5313
51. Bozomitu, R.G., Nita, L., et al.: A new integrated system for assistance in communicating with and telemonitoring severely disabled patients. Sensors **19**(9), 2026 (2019). https://doi.org/10.3390/s19092026
52. Hnatiuc, M., Caranica, A.: Communication between the sensor levels for monitoring subjects with disabilities. In: Proceedings of Advanced Technologies for Enhanced Quality of Life, pp. 87–91 (2009)
53. Hnatiuc, M., Belconde, A., Kratz, F.: Location of a person by means of sensors' network. In: Proceedings of ARTIPED 2010, pp. 301–305 (2010)
54. Schmitt, K.U., Muser, H., et al.: Comparing eye movements recorded by search coil and infrared eye tracking. J. Clin. Monitor. Comput. **21**, 49–53 (2007)
55. Kong, Y., et al.: Low-cost infrared video-oculography for measuring rapid eye movements. In: Park J., Chen, S.C., Raymond Choo, K.K. (eds.) Advanced Multimedia and Ubiquitous Engineering. FutureTech 2017, MUE 2017. Lecture Notes in Electrical Engineering, vol. 448. Springer, Singapore. Clerk Maxwell, J.: A Treatise on Electricity and Magnetism, vol. 2., 3rd edn. Clarendon, Oxford, 1892, pp. 68–73 (2017)
56. Johns, M.W., Tucker, A., et al.: Monitoring eye and eyelid movements by infrared reflectance oculography to measure drowsiness in drivers. Somnologie Schlafforschung Schlafmedizin **11**, 234–242 (2007)
57. Lupu, R., Ungureanu, F.: A survey of eye tracking methods and applications. Sci. Bull. Polytech. Inst. Autom. Control Comput. Sci. Sect. **3**, 71–86 (2013)
58. Duchowski, A.T.: Eye Tracking Methodology: Theory and Practice, Chapter 5: Eye Tracking Techniques, pp. 51–59. Springer (2007)
59. Chau, M., Betke, M.: Real time eye tracking and blink detection with USB cameras. Technical Reports, No. 12, pp. 1–10. Boston University Computer Science (2005)
60. Andrienko, G., Andrienko, N., Burch, M., Weiskopf, D.: Visual analyrics methodology for eye movement studies. IEEE Trans. Visual Comput. Graph. **18**(12), 2889–2898 (2012)

61. Barae, R., Boquete, L., Mazo, M.: System for assisted mobility using eye movements based on electrooculography. IEEE Trans. Neural Syst. Rehabil. Eng. **10**(4), 209–218 (2002)
62. Arthi, S.V., Norman, R.: Interface and control of appliances by the analysis of electrooculography signals. In: Artificial Intelligence and Evolutionary Computations in Engineering Systems, pp. 1075–1084. Springer, India (2016)
63. Wu, S.L., Liao, L., et al.: Controlling a human-computer interface system with a novel classification method that uses electrooculography signals. IEEE Trans. Biomed. Eng. **60**(8), 2133–2141 (2013)
64. Barea, R., Boquete, L., et al.: Wheelchair guidance strategies using EOG. J. Intell. Robot. Syst. **34**, 279–299 (2002)
65. Dhillon, H.S., Singla R., Rekhi, N.S., Jha, R.: EOG and EMG based virtual keyboard: a brain—computer interface. In: Proceedings of 2nd IEEE International Conference on Computer Science and Information Technology, pp. 259–262 (2009)
66. Usakli, A.B., Gurkan, S.: Design of a novel efficient human–computer interface: an electrooculagram based virtual keyboard. IEEE Trans. Instrum. Measure. **59**(8), 2099–2108 (2010)
67. Deng, L.Y., Hsu, C.L., et al.: EOG based human-computer interface system development. Expert Syst. Appl. **37**, 3337–3343 (2010)

Chapter 3
Intelligent Functional Electrical Stimulation

Marian-Silviu Poboroniuc and Dănuţ-Constantin Irimia

Abstract Functional Electrical Stimulation (FES) holds the premises to artificially control the musculoskeletal system aiming to improve quality of life in e.g. multiple sclerosis patients, or to provide targeted rehabilitation in e.g. stroke patients. Besides some neuromuscular stimulators which are widely used within FES clinics (e.g. Odstock Drop Foot Stimulator to correct foot drop in poststroke rehabilitation), some other FES-based control strategies e.g. to restore gait in paraplegia, are still under intensive research. The proposed chapter will shortly review the FES-based applications in neurorehabilitation and then will focus on current research that aims to artificially control the human body muscles by means of FES in order to, e.g. restore gait in paraplegia, improve neurorehabilitation in stroke patients, as well as the new trends to combine FES with hand and arm orthoses and Brain-Computer Interface (BCI).

Keywords Functional electrical stimulation · Neuroprostheses control · SCI and CVA rehabilitation · Brain-computer interfaces · Intelligent neuroprostheses

3.1 Introduction

There are already more than fifty years since the first attempts to design devices aiming to improve poststroke rehabilitation by means of electrical stimulation. This first electrical stimulator aimed to prevent foot-drop in hemiplegic patients [35]. Electrical stimulation is a technique of applying safe levels of electric current to generate artificially controlled muscular contractions and produce a useful movement. The devices integrating the programmable neurostimulators, sensors and specific control strategies are termed as neuroprostheses. The technique that uses neuroprostheses is known as Functional Electrical Stimulation (FES) and it may generate complex body movements in individuals whose muscles are paralysed due to an injury to the central nervous system (e.g. stroke (CVA), multiple sclerosis (MS), spinal cord

M.-S. Poboroniuc (✉) · D.-C. Irimia
"Gheorghe Asachi" Technical University of Iaşi, Iaşi, Romania
e-mail: mpobor@tuiasi.ro

H. Costin et al. (eds.), *Recent Advances in Intelligent Assistive Technologies: Paradigms and Applications*, Intelligent Systems Reference Library 170,
https://doi.org/10.1007/978-3-030-30817-9_3

injury (SCI) [4, 13, 47, 75]. The main objectives of applying FES depend on the disease. For example, FES-assisted therapy in stroke aims to increase force and decrease atrophy in paretic muscles, to inhibit spasticity, to improve proprioception through stimuli elicited in tendon and muscle receptors, to correct foot drop, and finally to lead to a normal life, or if still not possible, to improve the quality of life by using neuroprostheses etc. All FES-based rehabilitation techniques require a careful analysis of the patient's needs and specific disease treatment, the required sensorial system, specific electrostimulation electrodes and control strategy embedded within a neuroprosthesis. This chapter addresses the new FES-based rehabilitative devices personalized for specific users, specific needs in terms of used surface electrodes, and hybrid combination FES-robotic devices and BCI.

3.2 General Contexts, Potential Users and Statistics

The overall FES-based applications are part of a complex field of research and clinical rehabilitation procedures while they are intended to benefit users in multiple ways (e.g. improved walking in stroke [33, 35, 68, 74], upper limb rehabilitation after stroke [14, 39], FES as part of a rehabilitation methods in spinal cord injured (SCI) people [58], FES phrenic pacing system [27], BCI&FES to restore motor functions [15, 23, 24, 37], etc.). As one may observe there is a tremendous diversity of FES-based, and newly FES & BCI-based, applications tailored to each individual's rehabilitation needs and adapted for daily clinical use.

3.2.1 Potential Users

3.2.1.1 Stroke

Stroke (or Cardiovascular Accident (CVA)) is one of the leading causes to induce adult disability all over the world [69]. In accordance to WHO (World Health Organization) statistics studies, every year, more than 15 million people suffer stroke worldwide. Of these, 5 million are permanently disabled and have to undergo a rehabilitation programme. It has to be personalised for each individual and FES earned more and more application over the years to sustain such a rehabilitation process. For example, the inability to lift the foot has been addressed by electrical stimulation of the peroneal nerve during walking, the stimulation being triggered by a switch placed under the heel. Since 2009, the NHS(UK) recommended it as interventional procedure at national level [43]. Other interventional procedures address the upper limb rehabilitation [5, 18, 32]. A review found a statistically significant benefit from FES applied within 2 months of stroke on the primary outcome of ADL (Activities of Daily Living) [14]. Most of these new therapies which combine FES and robotic devices exploit the motor relearning principles. Electrical stimulation can be applied also to reduce spasticity, increase the muscle strength, improve the range of movement, and they are specifically adapted to each patient usually as a prerequisite for starting walking exercises, upper limb a-like daily exercises etc.

3.2.1.2 Multiple Sclerosis

Multiple Sclerosis (MS) is known as a chronic autoimmune demyelinating central nervous system disease. Much like the stroke, it is a leading cause of disability even in young adults. MS started to be diagnosed between 20 and 40 years old, and for example, 200 new MS cases are diagnosed each week in United States [42], and about the same rate is found in Northern European countries. In opposition to the stroke FES-based treatment outcomes which may lead to stroke patients' fully recovery, the use of FES in MS cases aims mostly to improve the life quality in users, e.g. by improving walking and reduce falls [2, 40, 71, 72]. FES may also be applied to reduce spasticity, to restore hand function and reduce muscle atrophy. MS-related tremor and trigeminal neuralgia may be reduced by techniques such as deep brain stimulation. MS-related pain and bladder dysfunction are treated by electrical stimulation at the spinal cord level. Bladder overactivity also responds to sacral neuromodulation and posterior tibial nerve stimulation [2]. Recently, some implantable FES devices started to be used as an alternative to the conventional transcutaneous stimulation to correct foot drop in MS users [71].

Several studies compared the orthotic and therapeutic effects of a foot drop stimulator on walking performance of subjects with chronic nonprogressive (e.g. CVA) and progressive (e.g. MS) disorders [68, 74]. One of them [68], involved 41 subjects with nonprogressive and 32 subjects with progressive conditions that used a foot drop stimulator for 3–12 months while walking in the community. The 10 m Walking Speed test with/without FES and PCI test were performed. The results showed that subjects with progressive and nonprogressive disorders had an orthotic benefit from FES up to 11 months, and the therapeutic effect increased for 11 months in nonprogressive disorders but only for 3 months in progressive disorders. The combined effect remained significant and clinically relevant.

3.2.1.3 Spinal Cord Injury (SCI)

In individuals suffering a traumatic injury of the spinal cord, a bundle of nerves as part of the central nervous system are affected too, damaging sensory, motor, and reflex capabilities below the injury site. The global prevalence of SCI is reported to be between 236 and 1009 per million [62] for different European countries and USA. USA reports between 245,000 and 353,000 living SCI in 2017, with an estimated of 17,500 new SCI each year, which e.g. costed the U.S. healthcare system more than $40.5 billion each year (since 2013, [1]). Depending of the spinal cord lesion the FES-based treatment may target different outcomes. For example, T7–T12 patients may benefit from FES-based neuroprostheses assisting the walking, and for many of them they perform transfer functions as wheel chair-toilet, wheel chair-bed, to improve their quality of life [47, 51]. FES had also positive effect to improve metabolism and induce positive trophic changes (e.g. the quadriceps femoris muscle fiber diameters showed an average increase of 59% after 8 months FES training) [28]. Some other SCI people, with high level spinal cord injury e.g. C5–C6, may benefit from using sophisticated FES systems to restore hand functions [18]. Freehand system [64]. Implanted FES-devices may help to improve bladder and bowel and respiration [66].

The above presented data refers to the most usual application of FES from the user's and professional's perspective, but under supervision of a specialist the procedure may be applied to other medical conditions, e.g. in scoliosis [30], the electrical stimuli being applied on the convex side of the spinal curvature.

3.3 Electrodes for Functional Electrical Stimulation

The electrical stimulation may be transcutaneous electrical nerve stimulation or stimulation directly on the targeted nerve by means of implanted electrodes. This chapter refers mostly to so-called enervated muscles in which contraction may be induced by stimulating the nerve which supplies the muscle. People suffering a lesion at the central nervous system level (CVA, MS, SCI, cerebral palsy) will experience difficulties to control the peripheral nervous system which remains intact (or partially damaged in MS) and can be controlled by locally providing the electrical stimuli to the peripheral muscle nerves. The usual programmable/non-programmable neurostimulators, aiming to produce tetanic muscle contractions, account for a tuning of electrical stimulation parameters as follows: asymmetrical or symmetrical biphasic voltage driven wave form, current up to 100 mA into a 1 kΩ load, 20–60 Hz frequency, 3–500 μs pulse width. Depending on the FES device, some other adjustment parameters that shape the wave of the electrical stimulus in order to adapt to patient's needs, may be provided (e.g. Odstock Medical Ltd. Foot drop stimulator ODFS Pace: ramping times 0 to 4 s, extension time: 1–1.5 s, output time: 0.5–6 s).

The main requirements for electrodes are to be biocompatible and safe. From the point of view of clinical practice and the ambulatory use the surface electrodes have to be easy in donning and doffing, to last for a long use and provide sufficient selectivity to guaranty that only the targeted motor system is activated. It is obvious that implanted electrodes provide better selectivity than surface electrodes [76]. Solutions to improve selectivity have been provided also with matrices of surface electrodes [6, 44]. O'Dwyer et al. [44], proposed an FES system for upper limb rehabilitation that uses instead of a single pair of surface electrodes, a normal pre-gelled electrode and the second electrode as a matrix of electrodes. This arrangement identifies the optimum hand response while selecting different second electrode's part from the matrix.

Most of the existing FES devices use pre-gelled electrodes and many times the patients complain that they have to find a better position for them anytime they want to use the FES device, e.g. the ODFS Pace for walking. Much more, the patients are not aware of the moment they have to replace them with new ones because they are not sticky anymore and a tube grip is used to keep them in place. Newly, our SCECM research group in Iasi developed the technology and know-how to integrate textile electrodes in trousers. These trousers with embedded knitted electrodes are obtained following a normal technological process on circular knitting machines. Their effectiveness has been tested in a clinical rehabilitation hospital in Iasi, Romania [48]. At most, in CVA patients, the FES-based rehabilitation exercises require the stimulation over the peroneal nerve to produce dorsiflexion, as well as the quadriceps contraction for several algorithms aiming to provide support in the stance phase. Once identified the required positions for the electrodes, different sizes (S, M, L) of trousers that embed textile electrodes were manufactured (see Fig. 3.1).

The main knitted part of the trousers was manufactured as a tubular jersey with a knitted yarn that contains 96% cotton and 4% elastane. After intensive testing, the Silver-coated electroconductive yarn, Shieldex® 117/17 dtex 2-ply HC + B (final fineness: 300 dtex, resistivity of 115 Ω/m) was chosen to knit the electrode isles. Besides the trousers presented in Fig. 3.1 new improvement was added, e.g. by covering the outside part of the electrodes with small pockets having inside small wet sponges (Fig. 3.2).

A group of five patients participated in the clinical trials after being instructed and providing informed consent. They used pre-gelled electrodes and later, the same day, they tested the NOVAFES trousers. The quantitative measurements take into account the 10 m walking speed with/without FES and Physiological Cost Index (PCI). The walking speed with FES with classical pre-gelled electrodes, as compared with no FES, improved by a mean of 12.09%. For the same patients, the same day and under the same conditions, the walking speed with FES with NOVAFES trousers,

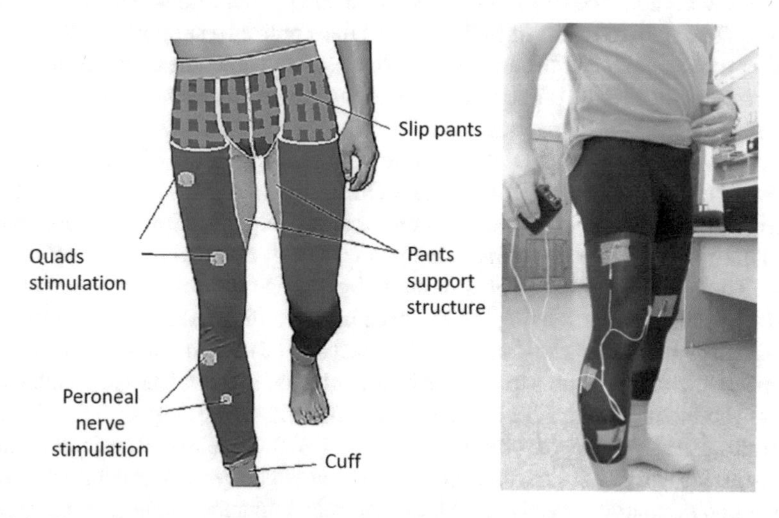

Fig. 3.1 Electrodes placement diagram of knitted NOVAFES trousers with embedded electrodes for peroneal nerve and quadriceps stimulation in stroke patients (left) [48]; The manufactured form of knitted trousers with embedded electrodes for walking in CVA patients (right)

Fig. 3.2 The clinical trials of the NOVAFES trousers with embedded electrodes to correct drop foot in stoke patients

as compared with no FES, improved by a mean slightly higher than the one when classical electrodes were used. In both cases the walking effort (PCI measurements) while using FES, compared to no FES, decreased by a mean over 20% (about 5% decrease in NOVAFES trousers). A comfort questionnaire measured the degree of patient satisfaction in using the NOVAFES trousers (e.g. ease in donning and doffing, thermal sensation, perceived humidity, tactile pleasantness, etc.). Overall the CVA patients well appreciated the new set-up that include the NOVAFES trousers to improve their walking [48].

3.4 Intelligent Functional Electrical Stimulation

Besides the compliance of the FES system to each patient treatment (e.g. in CVA and MS for improved gait, in SCI for muscular control to support transfer and/or walking movements) and the adequacy to the expected outcomes, proper placement of electrodes, other parts of the FES systems have to be properly designed (e.g. light sensorial system, control methods). The specific research literature outlines the idea that restoration of any function would have better chances if the FES-based systems mimic the biological control that has hierarchical structure [51, 77]. The highest artificial control levels take into account discrete control (e.g. selection of a programme), while the lower levels control include dynamics (e.g. model-based control at joint level). The entire body has to be monitored by a proper sensorial system, to detect also the user's intentions, being the interface between the user/artificial control system and the FES. The Brain-Computer Interfaces (BCI) are actually very promising sensorial systems being able to provide information on users' wishes at cortical level and translating them into FES-based controls at his/her lower/upper limbs [23, 24]. Controlling the gait of a SCI subject by using only FES and sensorial systems requires a tremendous amount of data as well as better human body models/submodels (e.g. human body dynamic models, FES-based muscle activation model), that currently impede their usefulness in a clinical environment or in ambulatory. Progress has been made in terms of sensorial systems while actual micro- and nano-technologies allow real-time monitoring of position, velocity, acceleration, space orientation, distance to an object, force detection in grasping etc. [36, 50, 57]. Some of the current sensors are highly miniaturized, require low energy and very important, can be embedded within a microcomputer and communicate wireless. Anyway, at the moment there is no sensorial system allowing for non-invasive measurement of the force generated by an individual muscle that might be required by some complex control strategies supporting standing in SCI (e.g. PDMR [54]). Electromyography (EMG) might be used to monitor muscular activity but it is difficult to perform it while applying electrical stimulation [34].

Only few FES-based devices aiming to support rehabilitation and daily activities outside the clinic are currently available and effective (e.g. ODFS Pace [74], WalkAide [53]). They have been kept as simple as possible in terms of required set-up, electrode placement, donning and doffing and sensors. For example, a foot drop stimulator (e.g. ODFS PACE) requires only two surface electrodes placement over the peroneal nerve on the affected side, placement of a switch (sometimes on the opposite side with a proper set-up of an internal device parameter) in the shoe under the heel and adjustment of the electrical stimulus (the main parameters, such as current intensity, ramping, period of stimulation if not triggered by the switch, symmetric/asymmetric wave forms, are adapted to each patient by a trained clinician). Other complex FES-based devices e.g. for walking in SCI still remain at clinical stage level with limited number of trials and use. Some FES-based control strategies (e.g. SCI standing-up, stance and gait) are of interest while the miniaturization of sensorial systems evolves and the use of biological sensors might become an alternative [63]. The information might be extracted by recording from peripheral nerves while these biological sensors send data to the central nervous system through afferent neurons.

The performance of neuroprostheses for SCI standing strongly depends on the control strategy and the reliability of the feedback signals. Each of standing subtasks (standing-up movement, standing and sitting-down movement) may require a specific control strategy and usually the controller embeds these and a parameter allowing manually/automated switching among them. Basically, there are two approaches: patient-centred and controller-centred strategies. Patient-centred strategies (e.g. PDMR [54], CHRELMS [12]) accounts for the voluntary contribution of the patient in making the control decision, while the controller-centred strategies (e.g. ONZOFF [47]; see Fig. 3.3b, [11]) On/Off ([41]; see Fig. 3.3a) need the patient's involvement in complying with the controller.

For example, within the CHRELMS (Control by Handle Reactions of Leg Muscle Stimulation [12]) control strategy, the arm forces as voluntary intentional input from the patient have to be measured and the required stimulation pattern is calculated to substitute the load by the lower limb while assuming a quasi-static movement. The required leg joint moments are calculated on a one-legged humanoid model basis. Additionally, CHRELMS requires a substantial set of sensors measuring handle forces and joint positions, which limit practical application.

Another FES-based control strategy, PDMR (Patient Driven Motion Reinforcement) [54] requires the joint positions and velocities measurement as feedback signals

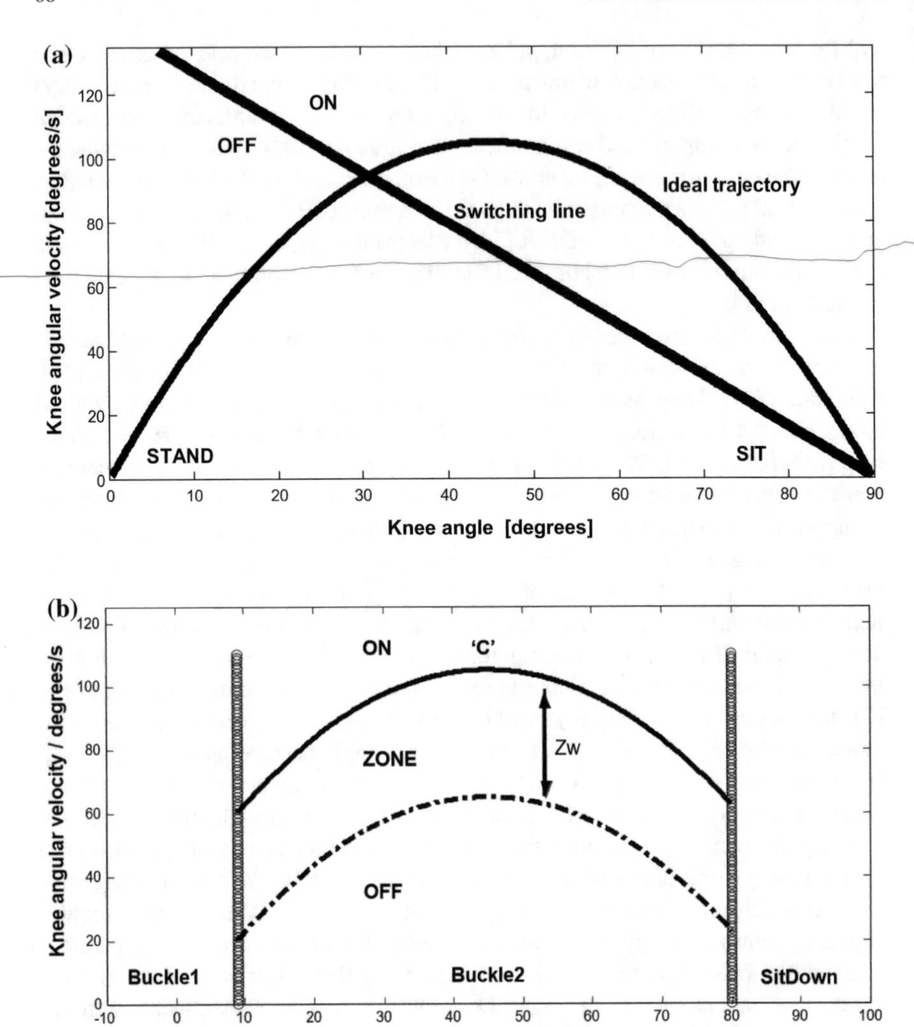

Fig. 3.3 **a** State-space diagram for knee angle against knee angular velocity as for the On/Off controller (adapted from Mulder et al. [41]); **b** state-space diagram for knee angle against knee angular velocity as for the ONZOFF controller (adapted from Poboroniuc et al. [47])

to an inverse dynamic model that calculate the stimulation (pulse width) needed to sustain the initiated movement.

The PDMR strategy does not require estimation of handle reaction. Compared to CHRELMS, this strategy requires fewer sensors since only the position and velocity of the body segments is needed, but it has only been experimentally validated for standing up and sitting down movements in a simplified situation with the patient sitting on a see-saw construction [55]. In addition, a major disadvantage of this controller is that awkward parameter adjustment is required prior to the experiments.

Both, CHRELMS and PDMR control strategies have limited practical application because of the high number of sensors required and the amount of time required in clinical assessment to measure or identify the high number of parameters.

The On/Off controller [41] supports standing-up or sitting-down in paraplegia and works by activating the quadriceps muscles in relation to a switching line in the state-space of knee angular velocity against knee angle (see Fig. 3.3a). For example, to support standing-up it provides maximum (ON; over the switching line) or minimum (OFF; under the switching line) quadriceps muscle activation. However, this controller was tested in experiments only with the patient in the supine position and, therefore, perturbations and hand force reactions of the patient were avoided.

A better control strategy ONZOFF [47] works according to a switching curve in knee angle against knee angular velocity state-space, but with a gradual increase or decrease in stimulation pulse width between the On and Off sub-spaces, the so-called 'Zone' (see Fig. 3.3b). The system requires only knee angle measurements as feedback and keeping it simple in respect of required number of sensors makes it particularly useful for regular use at home.

The overall system that sustain a chained motion *standing-up—standing—sitting-down* [47] is ease to be implemented within a programmable neurostimulator and requires only the knee angle measurement, being a good option for SCI subjects performing exercises or using it at home on a regular basis. For each subtask, the control strategy can be chosen as follows: standing-up (open loop control, On/Off, ONZOFF), standing (PID control [78], Knee Extension controller (KEC) [17]) and sitting-down (On/Off, ONZOFF) [47]. These type of neuroprostheses have been clinically tested and proved to be valuable for SCI patients (see Fig. 3.4).

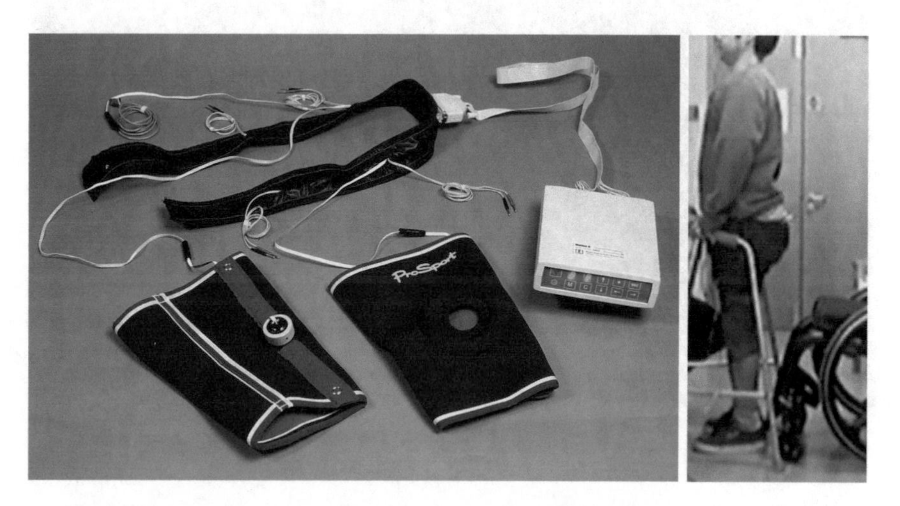

Fig. 3.4 Left—the standing supportive system for SCI individuals (knee cuffs with sensors, Stanmore programmable neurostimulator); right—clinical tests on a T7 SCI subject. Adapted from Poboroniuc [46]

The CVA and SCI people may also benefit from using FES to exercise the upper limb [14, 70]. Usually the SCI patients may strengthen their upper limb muscles and have assisted voluntary muscle control, while in CVA, FES may reinforce the voluntary control conducive to total recovery in some cases. The problems are related to muscle selectivity and electrode placement to induce proper upper limb spatial control (see Fig. 3.5).

Attempts to improve grasping capabilities in complete SCI subjects (C5/C6/C7 level), by means of the FreeHand system (surgically implanted neuroprosthesis) are mentioned in literature [29, 73]. It provides stimulation to the forearm muscles. Only few systems have been implanted and have not proven to be a feasible long term solution. Neuroprostheses that use transcutaneous stimulation still remain a better choice. For example, the NESS Handmaster device [65], is an external FES unit much like a splint, being placed over the muscles of the forearm. Anyway, the high cost prohibits their extensive use outside clinics. The development of control strategies for FES in the area of upper limb rehabilitation still remains at the *bench* stage of FES research. On the other side, the combination of FES and light exoskeletons and robotics, as well as FES and BCI, seems to be a better way to target rehabilitation in CVA and SCI patients, and will be discussed within next subchapters.

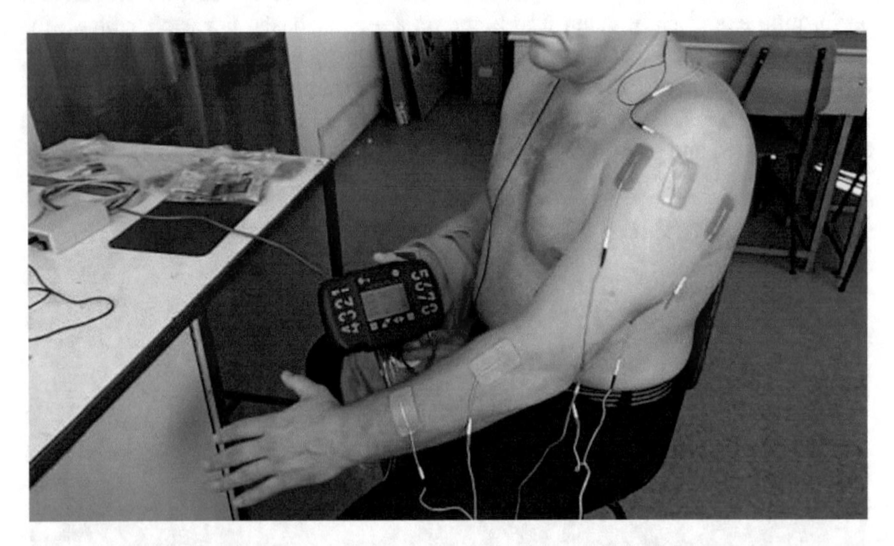

Fig. 3.5 Electrode placement over the shoulder (anterior-posterior deltoid and supraspinatus), triceps to support elbow extension and forearm to support wrist and fingers extension

3.5 Functional Electrical Stimulation and Robotic Technology

Functional electrical stimulation and robotic exoskeletons are two technologies which started to be widely used to support rehabilitation in CVA, SCI, CP patients [8, 9, 31]. Anyone of them brings positive outcomes but their combination has greater potential to impact the rehabilitation process. For example, FES activate the patient muscles in a coordinated way inducing cortical reorganization while the mechatronic equipment sustains the optimal movement of the upper limb/lower limb as well as provides quantitative evaluation of the rehabilitative process. Much more, the rehabilitation robotics represents a tremendous support for clinicians by enhancing their productivity and effectiveness in their effort to sustain the CVA, SCI, CP individual's recovery.

Firstly, gait orthotics benefited more from using mechatronic devices. The so-called walking orthotics aims to sustain a stable weight bearing, to control the speed/direction of lower limb motion, and to reduce the patient's energy consumption during walking ([16], HAL® Lower Limb as mobile medical rehabilitation exoskeleton (Cyberdyne Inc.), ReWalk from ReWalk Robotics Ltd., Lokomat from Hocoma, G-EO system from REHA Technology). These powered exoskeletons still have to prove their effectiveness in rehabilitation and not only as walking assistive devices. Some of these devices provide walking assisted support on a treadmill (e.g. Lokomat from Hocoma), and as far as an incomplete SCI progress he/she may use a free walking support system (e.g. Andago from Hocoma). There is a lack of published studies to confirm their effectiveness in different environments (e.g. slopes, different surfaces), but anyway, they started to become a viable alternative to wheelchairs.

A great interest reached the so-called hybrid exoskeletons which combine FES and active exoskeletons. Some of the drawbacks of each individual system (muscle fatigue, joint trajectories control difficulties) which limited their wide spreading in clinical practice, can be overpassed. Of the actuality are the fully active hybrid exoskeletons that can dissipate and deliver power to the joint. Therefore, the insufficient muscular response in neurologically injured subjects and the muscle fatigue due to stimulation can be compensated. A drawback is that the fully-active systems are energetically inefficient and bulky. For now, the clinical application of these devices (e.g. GE-O, Lokomat) allow only some predefined pattern (time to ON and OFF the electrical stimuli over the selected group of muscles) of electrical stimulation in accordance with walking. A further improvement has to take into account a closed-loop control of FES based on indirect measures of muscle performance.

The rehabilitation robotics and FES for upper limbs received great interest too (e.g. MIT-Manus robot, Armotion from Reha Technology, H200 wireless rehabilitation system from Bioness, Armeo®Spring from Hocoma, Amadeo from Tyromotion, [18, 31, 67]). A recent meta-analysis [21], focused on rehabilitation of the upper limb motor function after stroke and compared different techniques contributing to upper limb recovery. Based on a sufficient number of subjects (more than 500) as evidence, the authors recommend passive electrical stimulation (the so-called active one can be triggered by EMG activity) as an adjuvant therapy into stroke rehabilitation strategies aiming to improving upper extremity motor impairments. Mental practice with motor imagery further benefit the patient in the subacute and chronic post-stroke phase. Most of the mechatronic devices are designed to support the elbow and shoulder actuation and there is a lack of robotic training devices for finger and wrist movements. The upper limb robotic devices may be passive systems which only stabilize the upper limb, active systems which move the limb and the top research ones are interactive systems which allow the voluntary action of the patient into the control strategy of the targeted task [56]. Task-oriented training with robotic devices is frequently interfaced with virtual reality presented on a computer screen as a technique that better motivate the user to perform the exercises [26, 37].

Assisting robotics allows patients to perform repeated movements in a well-controlled manner, while the FES-based activation of the muscles helps the brain to relearn daily movements. This phenomenon has been termed as neuroplasticity. Some of our findings with a novel hybrid FES exoskeleton EXOSLIM system sustain that hypothesis [59]. The novelty of the proposed EXOSLIM device consists in a balanced control of the upper limb movements induced by the driven exoskeleton and the FES-based muscle activation (see Fig. 3.6). The main part of the upper limb rehabilitation system consists of a mechanical exoskeleton type structure, anthropometrically sized, aiming to ensure basic anatomic movements (shoulder flexion-extension, shoulder abduction-adduction, shoulder medial rotation and forearm flexion-extension) together with functional electrical stimulation. It has a modular, reconfigurable structure, thus adaptable for either the right or the left sides. The system has been successfully tested clinically on two subjects which were satisfied with the design and performed exercises. More clinical tests are under run. Much more, the EXOSLIM system has been combined with BCI (motor imagery tasks) and it proved the feasibility to control the EXOSLIM system via BCI. As requested by a visual paradigm, the user had to imagine either the right or left hand movement. If the imagined movement was classified as the one requested by the paradigm, the entire

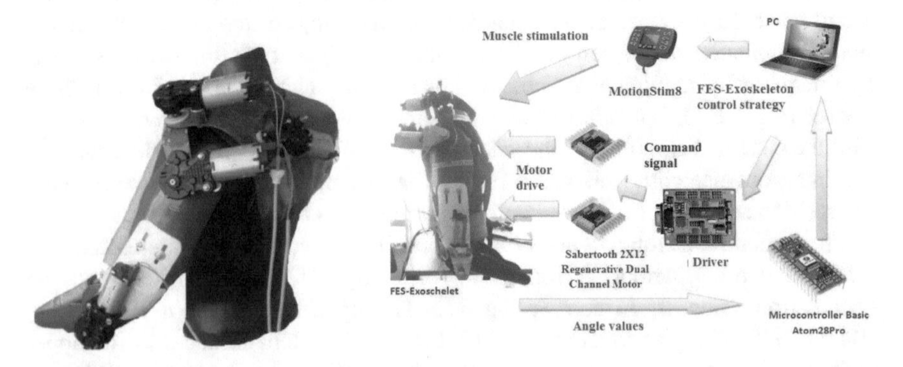

Fig. 3.6 The upper limb exoskeleton EXOSLIM and the general scheme of the control structure. Adapted from Serea et al. [59], Poboroniuc et al. [49]

upper limb performed a predefined spatial motion (FES & exoskeleton controlled) or relaxed based on the one of the two detected classes [25].

Another rehabilitative system IHRG targeted the hand and its fine control. The proposed hand rehabilitative IHRG system supports a balanced control between FES and an actuated glove while taking into account the patient's intention to perform hand/fingers opening movements (see Fig. 3.7) [20, 49].

A clinical randomized controlled study was conducted to test the IHRG system [18]. The IHRG-based therapy has been followed over 12 sessions, each lasting 45 min with a total of 9 h for each patient for a total number of 13 patients. The control group (12 patients) underwent conventional therapy in 10 sessions of 30 min each. The patients in the IHRG group showed a slight increase of average motor gain than the control group assessed with Fugl-Meyer Assessment (FMA) and Box and Blocks test (BBT). As compared to other studies which extended the therapy over more than five weeks, due to the Rehabilitation Hospital internal regulations in that study the patients were able to perform the therapy only for two weeks as long as they were admitted in the hospital facilities. The clinical trials with IHRG system are still running on. As a conclusion, better outcomes in CVA, SCI, CP rehabilitation will be obtained while adopting hybrid FES & mechatronics devices as rehabilitative devices, as well as completed with BCI systems, which will be commented further on.

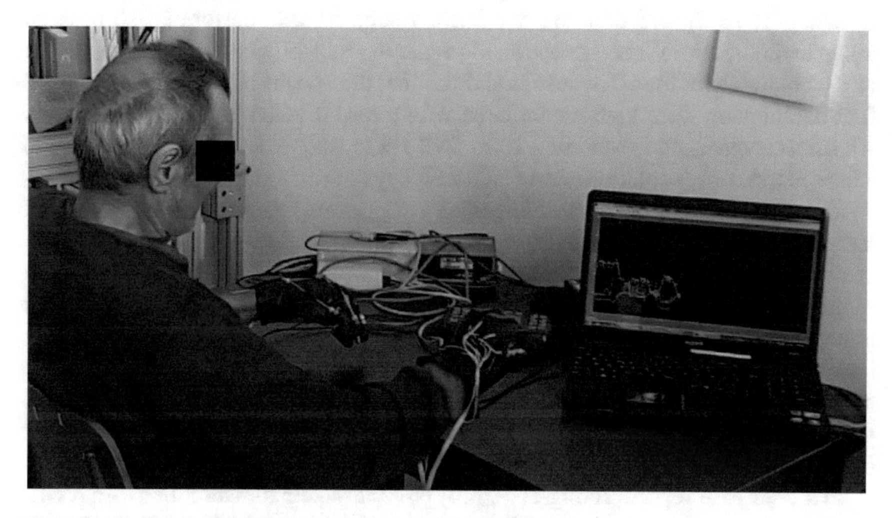

Fig. 3.7 Clinical tests and design set-up of the IHRG system

3.6 Functional Electrical Stimulation & Brain-Computer Interface

In the last years one of the newest investigated therapies, the BCI, has as main goal the upper/lower limb motor improvement in stroke patients. During the conventional rehabilitation therapy, the CVA patients are instructed to try to perform movements with the paretic limb, or imagine its movement, while a therapist, robotic device or electrical stimulator helps the patient in performing the desired movement. One of the main disadvantages of the conventional therapy is that there is no objective way to determine if the patient is actively performing the desired motor imagery (MI) task and the feedback may be provided when the user is not performing the required mental activity, fact that produces nonconcordant neural activation. The BCI technology solves this problem by providing an objective tool for measuring the motor imagery and therefore creating new possibilities for "closed-loop" feedback [23, 24].

Furthermore, many publications provide evidence that using motor imagery-based BCIs can induce neural plasticity and thus serve as important tool for the rehabilitation process of stroke patients [3, 19, 52, 60]. In [10], Do et al. tested the feasibility of integrating a neurostimulator with non-invasive BCI. The neurostimulator was used to elicit the foot dorsiflexion by transcutaneous stimulating the tibial anterior muscle. The study was done on five healthy subjects who performed 10 trials of idling and repetitive foot dorsiflexion to trigger the BCI-FES controlled dorsiflexion of the contralateral foot. The results showed that the epochs of BCI-FES controlled foot dorsiflexion were highly correlated with those of voluntary foot dorsiflexion. All subjects managed to achieve a 100% BCI-FES response and a single subject had a false alarm. Daly et al. [7] tested a BCI&FES system on a stroke patient presenting dyscoordination of isolated index finger joint extension of the metacarpal phalangeal joint. The experiment consisted of trials in which the patient had to attempt a finger movement, resting trials and trials where the user had to imagine the finger movement. During the first session, the subject had a good control accuracy for attempted movement (97%), imagined movement (83%) and some difficulties with attempted relaxation (65%). During the sixth session, the accuracy of controlling the relaxation improved to more than 80%. After nine sessions (three weeks' time), the researchers concluded that the patient's volitional isolated index finger extension has been improved.

In a later case report, Young et al. [80] present the combination between a BCI system with coordinated visual, FES and tongue stimulation (TDU) feedback modalities designed to improve the rehabilitation process results for the upper limbs of a subacute stroke patient. The participant underwent 6 weeks of interventional rehabilitation therapy using BCI-FES-TDU system. The assessment consisted of anatomical and functional MRI scans, while the behavioural measures included the Stroke Impact Scale (SIS) and the Action Research Arm Test (ARAT). Clinically significant improvements were observed, with more than 10 point gains in both ARAT and

SIS hand function domain scores. The neuroimaging during finger tapping of the impaired hand also showed changes in the brain activation patterns associated with the BCI therapy.

Since 2012, one of our main research interest, in collaboration with a company producing BCIs (g.tec medical engineering GmbH), was to combine the BCI and FES technologies in order to achieve improved results of the rehabilitation process for upper limb in stroke patients. Figure 3.8 presents the schematic illustration of the BCI&FES concept. In there, the user has to imagine or to perform specific movements, as for example the wrist dorsiflexion. The resulting electroencephalogram (EEG) activity is detected through a set of electrodes overlying the sensory motor areas of the brain, then being sent to an amplifier. After amplification and digitization, the signals are sent to a computer that performs the data analysis and controls the feedback 2D presentation on a computer screen. Like in conventional therapy, the users have to perform motor imagery and to receive feedback (specifically, visual feedback on a computer screen and FES feedback as artificially induced movements). Unlike conventional therapy, BCI&FES users also wear an EEG cap that monitors motor imagery that influences the feedback. The key element of this concept is the real-time connection between the brain activity and feedback. The system provides visual and FES feedback to the user only if he/she correctly imagines the left or right hand movement. Therefore, unlike conventional therapy, the feedback is always paired with the brain activity.

Before starting the clinical tests on patients, we supposed that patients will have difficulties in achieving high control accuracy of the BCI&FES system. We found the opposite. In [23, 24], all three patients started with accuracies above 80%. As

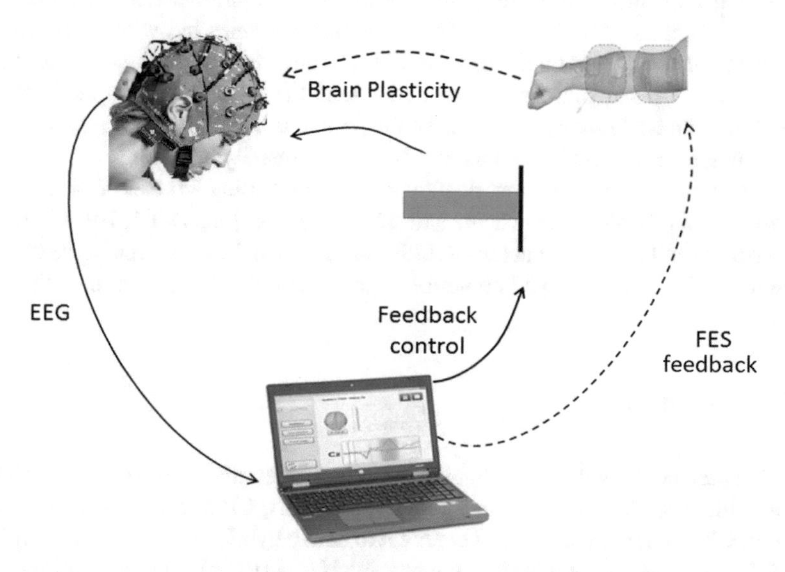

Fig. 3.8 The schematic view of the BCI&FES system (the head cap is a gNautilus device, GTEC, Austria)

averaged accuracies over all sessions, patient 1 reached 90.5%, patient 2 reached 85.4% and patient 3 reached 87.1%. Each patient managed to achieve an accuracy above 96.2% in at least one session. The sessions where the accuracies were low in each patient case are highly correlated with the patient's health condition during that day, emotional state and/or degree of tiredness. The motor improvement for patients 1 and 3 was assessed using the 9-hole PEG test, a game where the user has to fill 9 holes of a board with sticks placed on the table, and then to put the sticks back on the table. The test has to be performed for both hands and the time for accomplishing the task and the number of dropped sticks represent the score of that evaluation. Both improved the time they completed the task with the paretic hand with 35, respectively 38 s, while the time exercise for the healthy hands remained relatively constant in both cases. Patient 2 was not able to perform the 9-hole PEG test. He started to move the thumb after the 12th session and during the last sessions of training he started to perform small range voluntary movements also with the other fingers.

We found that also chronic stroke patients who did not benefit from conventional therapy during the study could improve motor function and BCI control accuracy by using a BCI&FES system for rehabilitation. We presented the results of 2 chronic stroke patients in a case study in 2017 [23, 24].

Patient P1, a 40-years old woman began the intervention 5.5 years after the stroke onset and had severe paralysis in her left hand with no residual movement. She had received conventional therapy for 2 years and no significant functional improvement had been observed before her participation in this study. The second patient, a 59 years' male began the intervention 3.25 years after the stroke. His left arm was almost 100% paralyzed, being able to move the middle finger in a range of 0.5 cm. Before participating to this study, in addition to conventional rehabilitation therapy, he performed TMS and mirror rehabilitation therapies but with no functional improvement. After 10 training sessions, both patients showed improved motor function. P1 was able to voluntarily relax and extend the wrist of her paretic hand. P2 started to voluntarily move all fingers of the paretic arm on a small range, while the middle finger movement range increased to approximately 1.5 cm.

These results extend prior work showing that BCI using MI can be an effective tool for motor rehabilitation in acute and subacute stages [38, 45, 61, 79]. Moreover, the current results suggest that the BCI&FES system can be used even in the chronic stage for patients who showed no motor improvement during conventional therapy.

3.7 Conclusions

This chapter deals with new rehabilitative techniques that aim to support recovery and increase the quality of life in CVA, MS, SCI, CP patients. It mainly starts with FES-based rehabilitation and extend it to newly hybrid techniques that involves rehabilitation robotics and brain-computer interfaces (BCIs). First part outlines the general context with definitions, electrical stimulation parameters and electrodes, patients that may benefit from using the presented rehabilitation techniques and then the exposal goes deep into the so-called *intelligent FES* by presenting the state-of-the-art, the combination FES-robotics-BCI and examples of the new designed rehabilitating systems that have been implemented in clinical practice (e.g. RecoveriX [22]) and are the very new outcomes of the performed research.

The FES-based rehabilitation technique is a constant presence within rehabilitation clinics and in some cases has been recommended as interventional procedure at national level (e.g. foot drop correction in CVA patients [43]). Anyway, we are far from producing a great difference for impaired people due to a lesion of central nervous system origin, as compared with traditional rehabilitation procedures, but it has the seeds to do it in the future. The success relies on better controllers to control the FES-based system, implanted or better surface electrodes, miniaturized sensorial system and even integration of the biological sensors within the designed control strategies. Furthermore, the interactive systems which allow the use of FES and mechatronic devices in order to integrate patient's wishes into the control strategy, in order to boost rehabilitation, have to be better designed. BCI comes to complete the rehabilitation schema by reconnecting the body parts to the brain via the provided feedback and its designed paradigms.

References

1. 2017 Spinal Cord Injury Statistics: On-line reference (2018). https://www.spinalcord.com. Accessed on 11 April 2018
2. Abboud, H., Hill, E., Siddiqui, J., Serra, A., Walter, B.: Neuromodulation in multiple sclerosis. Multiple Sclerosis J. **23**(13), 1663–1676 (2017)
3. Ang, K.K., Guan, C., Chua, K.S.G., Ang, B.T., Kuah, C., Wang, C., et al.: A clinical study of motor imagerybased brain-computer interface for upper limb robotic rehabilitation. Conf. Proc. IEEE Eng. Med. Biol. Soc. **2009**, 5981–5984 (2009)
4. Balasubramanian, S., He, J.P.: Adaptive control of a wearable exoskeleton for upper-extremity neurorehabilitation. Appl. Bion. Biomech. **9**(1), 99–115 (2012). https://doi.org/10.3233/ABB-2011-0041
5. Bissolotti, L., Villafane, J.H., Gaffurini, P., Orizio, C., Valdes, K., Negrini, S.: Changes in skeletal muscle perfusion and spasticity in patient with poststroke hemiparesis treated by robotic assistance (Gloreha) of the hand. Phys. Ther. Sci. **28**, 769–773 (2016)
6. Côté, M.: PA: Mapping of the human upper arm muscle activity with an electrode matrix. Electromyogr. Clin. Neurophysiol. **40**(4), 215–223 (2000)
7. Daly, J.J., Cheng, R., Rogers, J., Litinas, K., Hrovat, K., Dohring, M.: Feasibility of a new application of noninvasive brain-computer interface (BCI): A case study of training for recovery of volitional motor control after stroke. J. Neurol. Phys. Therapy **33**, 203–2011 (2009)
8. del-Ama, A.J., Gil-Agudo, Á., Pons, J., Moreno, J.: Hybrid FES-robot cooperative control of ambulatory gait rehabilitation exoskeleton. J. Neuroeng. Rehabil. **11**, 27 (2014)
9. Dingguo, Z., Yong, R., Kai, G., Jie, J., Wendong, X.: Cooperative control for a hybrid rehabilitation system combining functional electrical stimulation and robotic exoskeleton. Front. Neurosci. **11**, 725 (2017)
10. Do, A.H., Wang, P.T., King, C.E., Abiri, A., Nenadic, Z.: Brain-Computer Interface controlled functional electrical stimulation system for ankle movement. J. NeuroEng. Rehabil. **8**, 49 (2011)
11. Dolan, M., Andrews, B., Veltink, P.H.: Switching curve controller for FES assisted standing up and sitting down. IEEE Trans. Rehabil. Eng. **6**, 167–171 (1998)

12. Donaldson, N.N., Yu, C.H.: FES standing control by handle reactions of leg muscle stimulation (CHRELMS). IEEE Trans. Rehabil. Eng. **4**, 280–284 (1996)
13. Downey, R.J., Cheng, T.H., Bellman, M.J., Dixon, W.E.: Switched tracking control of the lower limb during asynchronous neuromuscular electrical stimulation: theory and experiments. IEEE Trans. Cybern. **47**(5), 1251–1262 (2017). https://doi.org/10.1109/TCYB.2016.2543699
14. Eraife, J., Clark, W., France, B., Desando, S., Moore, D.: Effectiveness of upper limb functional electrical stimulation after stroke for the improvement of activities of daily living and motor function: a systematic review and meta-analysis. Syst. Rev. **6**, 40 (2017). https://doi.org/10.1186/s13643-017-0435-5
15. Ethiera, C., Miller, L.: Brain-controlled muscle stimulation for the restoration of motor function. Neurobiol. Dis. **83**(180–190), 2015 (2015). https://doi.org/10.1016/j.nbd.2014.10.014
16. Fatone, S.: A review of the literature pertaining to KAFOs and HKAFOs for ambulation. J. Prosthet. Orthot. **18**(3), 137–168 (2006)
17. Fuhr, T., Quintern, J., Riener, R., Schmidt, G.: Assisting locomotion in patients with paraplegia. Control of WALK!—a cooperative patient driven neuroprosthetic system. IEEE EMBS Mag. **27**, 38–48 (2008)
18. Grigoras, V.-A., Irimia, D.C., Poboroniuc, M.S., Popescu, C.D.: Testing of a hybrid FES-robot assisted hand motor training program in sub-acute stroke survivors. Adv. Electr. Comput. Eng. **16**(4), 89–94 (2016). ISSN: 1582-7445, e-ISSN: 1844-7600, https://doi.org/10.4316/aece.2016.04014
19. Grosse-Wentrup, M., Mattia, D., Oweiss, K.: Using brain–computer interfaces to induce neural plasticity and restore function. J. Neural Eng. **8**(2), 025004 (2011)
20. Hartopanu, S., Poboroniuc, M.S., Serea, F., Irimia, D.C., Livint, G.: New issues on FES and robotic glove device to improve the hand rehabilitation in stroke patients. In: Proceedings of 6th International Conference on Modern Power System 2015. Acta Electrotehnica **56**(3), 123–127 (2015). ISSN: 1841-3323, ISSN: 2344-5637
21. Hatem, S., Saussez, G., della Faille, M., Prist, V., Zhang, X.: Rehabilitation of motor function after stroke: a multiple systematic review focused on techniques to stimulate upper extremity recovery. Front. Human Neurosci. **10**, 442 (2016)
22. Irimia, D., Sabathiel, N., Ortner, R., Poboroniuc, M., Coon, W., Allison, B.Z., Guger, C.: recoveriX: a new BCI-based technology for persons with stroke. In: 38th Annual International Conference of the IEEE Engineering in Medicine and Biology Society (EMBC); 08/2016, p. 1 (2016)
23. Irimia, D.C., Cho, W., Ortner, R., Allison, B.Z., Ignat, B.E., et al.: Brain-computer interfaces with multi-sensory feedback for stroke rehabilitation: a case study. Artif. Organs **41**, E178–E184 (2017). https://doi.org/10.1111/aor.13054
24. Irimia, D.C., Poboroniuc, M.S., Ortner, R., Allison, B.Z., Guger, C.: Preliminary results of testing a BCIcontrolled FES system for post-stroke rehabilitation. In: Proceedings of the 7th Graz Brain-Computer Interface Conference, Graz, Austria, 18–22 Sept 2017
25. Irimia, D,, Poboroniuc, M., Serea, F., Baciu, A., Olaru, R.: Controlling a FES-EXOSKELETON rehabilitation system by means of brain-computer interface. In: International Conference and Exposition on Electrical and Power Engineering, pp. 352–355 (2016). https://doi.org/10.1109/icepe.2016.7781361
26. Jang, S.H., You, S.H., Hallett, M., Cho, Y.W., Park, C.M., et al.: Cortical reorganization and associated functional motor recovery after virtual reality in patients with chronic stroke: an experimenter-blind preliminary study. Arch. Phys. Med. Rehabil. **86**, 2218–2223 (2005). https://doi.org/10.1016/j.apmr.2005.04.015
27. Jarosz, R., Littlepage, M., Creasey, G., McKenna, S.: Functional electrical stimulation in spinal cord injury respiratory care. Top Spinal Cord Inj. Rehabil. **18**(4), 315–321 (2012). https://doi.org/10.1310/sci1804-315
28. Kern, H.: Electrical stimulation on paraplegic patients. Eur. J. Trans. Myol. Basic Appl. Myol. **24**(2), 75157 (2014)

29. Kilgore, K.L., Peckham, H., Keith, M.W., et al.: An implanted upper-extremity neuroprosthesis. J. Bone Joint Surg. **79**, 533–541 (1997)
30. Ko, E.J., Sun, I.Y., Yun, G.J., Kang, J., Kim, J.Y., Kim, G.E.: Effects of lateral electrical surface stimulation on scoliosis in children with severe cerebral palsy: a pilot study. Disability Rehabil. **40**(2), 192–198 (2018)
31. Krebs, H.I., Volpe, B.T.: Rehabilitation robotics. Handb. Clin. Neurol. **110**, 283–294 (2013). https://doi.org/10.1016/b978-0444-52901-5.00023-x
32. Kutlu, M., Freeman, C.T., Hallewell, E., Hughes, A.-M., Laila, D.S.: Upper-limb stroke rehabilitation using electrode-array based functional electrical stimulation with sensing and control innovations. Med. Eng. Phys. **38**, 366–379 (2016)
33. Kyung-Hoon, Y., Kwon-Young, K.: Functional electrical stimulation with augmented feedback training improves gait and functional performance in individuals with chronic stroke: a randomized controlled trial. J. Kor. Phys. Ther. **29**(2), 74–79 (2017). https://doi.org/10.18857/jkpt.2017.29.2.74
34. Li, Z., Guiraud, D., Andreu, D., Benoussaad, M., Fattal, C., Hayashibe, M.: Real-time estimation of FES-induced joint torque with evoked EMG. Application to spinal cord injured patients. J. Neuroeng Rehabil. **13**, 60 (2016)
35. Liberson, W.T., Holmquest, H.J., Scott, D., et al.: Functional electrotherapy: stimulation of the peroneal nerve synchronized with the swing phase of the gait of hemiplegic patients. Arch. Phys. Med. Rehabil. **42**, 101 (1961)
36. Liu, Y., Wang, H., Zhao, W., Zhang, M., Hongbo, Q., et al.: Flexible, stretchable sensors for wearable health monitoring: sensing mechanisms, materials, fabrication strategies and features. Sensors **18**(2), 645 (2018)
37. Lupu, R.G., Irimia, D.C., Ungureanu, F., Poboroniuc, M.S., Moldoveanu, A.: BCI and FES based therapy for stroke rehabilitation using VR facilities. Hindawi Wireless Commun. Mobile Comput. **2018**, 8 (2018). Article ID 4798359. https://doi.org/10.1155/2018/4798359
38. Mattia, D., Pichiorri, F., Molinari, M., Rupp, R.: Part II: devices, applications and users brain computer interface for hand motor function restoration and rehabilitation. In: Allison, Z.B., Dunne, S., Leeb, R., Del, J., Millan, R., Nijholt, A. (Eds.) Towards Practical Brain-Computer Interfaces: Bridging the Gap from Research to Real-World Applications, pp. 131–53. Springer, Berlin (2013)
39. Mazzoleni, S., Duret, C., Gaëlle, A., Grosmaire, E.B.: Combining upper limb robotic rehabilitation with other therapeutic approaches after stroke: current status, rationale and challenges. Biomed. Res. Int. **2017**, 8905637 (2017). https://doi.org/10.1155/2017/8905637
40. Miller, L., McFadyen, A., Lord, A., Hunter, R., Paul, L., Rafferty, D., Bowers, R., Mattison, P.: Functional electrical stimulation for foot drop in multiple sclerosis: a systematic review and meta-analysis of the effect on gait speed. Arch. Phys. Med. Rehabil. **98**(7), 1435–1452 (2017)
41. Mulder, A.J., Veltink, P.H., Boom, H.B.K.: On/off control in FES-induced standing up: a model study and experiments. Med. Biol. Eng. Comput. **30**, 205–212 (1992)
42. Multiple Sclerosis by the Numbers: Facts, statistics, and you, on-line reference (2018). https://www.healthline.com/health/multiple-sclerosis/facts-statistics-infographic#1. Accessed on 11 April 2018
43. NHS (UK), Interventional procedure guidance 278: Functional electrical stimulation for drop foot of central neurological origin. National Institute for Health and Clinical Excellence (NHS) (2009). ISBN 1-84629-846-6
44. O'Dwyer, S.B., O'Keeffe, D.T., Coote, S., Lyons, G.M.: An electrode configuration technique using an electrode matrix arrangement for FES-based upper arm rehabilitation systems. Med. Eng. Phys. **28**, 166–176 (2006)
45. Ortner, R., Irimia, D.C., Scharinger, J., Guger, C.: A motor imagery based brain-computer interface for stroke rehabilitation. Stud. Health Technol. Inform. **181**, 319–323 (2012)
46. Poboroniuc, M.: Current status and future prospects for FES-based control of standing and walking in paraplegia. In: 3rd International Conference on Electrical and Power Engineering EPE2004, Bulletin of the Polytechnic Institute of Iasi, tom L (LIV), Fasc.5A, Iasi, Romania, pp. 21–32, 7–8 Oct 2004. ISSN 1223-8139

47. Poboroniuc, M., Wood, D.E., Riener, R., Donaldson, N.N.: A new controller for FES-assisted sitting down in paraplegia. Adv. Electr. Comput. Eng. **10**(4), 9–16 (2010). https://doi.org/10.4316/aece.2010.04002
48. Poboroniuc, M.S., Irimia, D.C., Poboroniuc, I.C., Curteza, A., Macovei, L., et al.: Manufacturing and clinically testing embedded electrodes in knitted textiles for neurorehabilitation. In: Proceedings of 2017 International Conference on Electromechanical and Power Systems (SIELMEN), pp. 68–73 (2017). https://doi.org/10.1109/sielmen.2017.8123294
49. Poboroniuc, M.S., Irimia, D.C.: FES&BCI based rehabilitation engineered equipment: clinical tests and perspectives. In: E-Health and Bioengineering Conference (EHB), pp 1–6 (2017). https://doi.org/10.1109/ehb.2017.7995365
50. Poegge, L.S., Tosi, D., Duraibabu, D.B., Leen, G., McGrath, D., Lewis, E.: Optical fibre pressure sensors in medical applications. Sensors **15**(7), 17115–17148 (2015)
51. Popović, D.: Advances in functional electrical stimulation (FES). J. Electromyogr. Kinesiol. **24**(6), 795–802 (2014). https://doi.org/10.1016/j.jelekin.2014.09.008
52. Prasad, G., Herman, P., Coyle, D., McDonough, S., Crosbie, J.: Applying a brain-computer interface to support motor imagery practice in people with stroke for upper limb recovery: a feasibility study. J. Neuroeng. Rehabil. **7**, 60 (2010)
53. Prosser, L., Curatalo, L.A., Alter, K.E., Damiano, D.L.: Acceptability and potential effectiveness of a foot drop stimulator in children and adolescents with cerebral palsy. Dev. Med. Child Neurol. **54**(11), 1044–1049 (2013)
54. Riener, R., Fuhr, T.: Patient-driven control of FES-supported standing up: a simulation study. IEEE Trans. Rehabil. Eng. **6**, 113–124 (1998)
55. Riener, R., Ferrarin, M., Pavan, E., Frigo, C.: Patient-driven control of FES-supported standing up and sitting down: experimental results. IEEE Trans. Rehabil. Eng. **8**, 523–529 (2000)
56. Riener, R., Nef, T., Colombo, G.: Robot-aided neurorehabilitation of the upper extremities. Med. Biol. Eng. Comput. **43**, 2–10 (2005). https://doi.org/10.1007/BF02345116
57. Sabatini, A.M.: Estimating three-dimensional orientation of human body parts by inertial/magnetic sensing. Sensors **11**(2), 1489–1525 (2011)
58. Sadowsky, C., Edward, R., Strohl, H., Commean, A., Eby, P., et al.: Lower extremity functional electrical stimulation cycling promotes physical and functional recovery in chronic spinal cord injury. J. Spinal Cord Med. **36**(6), 623–631 (2013). https://doi.org/10.1179/2045772313y.0000000101
59. Serea, F., Poboroniuc, M.S., Hartopanu, S., Irimia, D.: Towards clinical implementation of an FES&Exoskeleton to rehabilitate the upper limb in disabled patients. In: Proceedings of International Conference on Control Systems and Computer Science (CSCS), pp. 827–832 (2015). https://doi.org/10.1109/cscs.2015.114
60. Shindo, K., Kawashima, K., Ushiba, J., Ota, N., Ito, M., Ota, T., et al.: Effects of neurofeedback training with an electroencephalogram-based brain-computer interface for hand paralysis in patients with chronic stroke: a preliminary case series study. J. Rehabil. Med. **43**(10), 951–957 (2011)
61. Silvoni, S., Ramos-Murguialday, A., Cavinato, M., et al.: Braincomputer interface in stroke: a review of progress. Clin. EEG Neurosci. **42**, 245–252 (2011)
62. Singh, A., Tetreault, L., Kalsi-Ryan, S., Nouri, A., Fehlings, M.: Global prevalence and incidence of traumatic spinal cord injury. Clin. Epidemiol. **6**, 309–331 (2014)
63. Sinkjær, T., Haugland, M., Inmann, A., Hansen, M., Nielsen, D.K.: Biopotentials as command and feedback signals in functional electrical stimulation systems. Med. Eng. Phys. **25**(1), 29–40 (2003)
64. Smith, B.T., Mulcahey, M.J., Betz, R.R.: An implantable upper extremity neuroprosthesis in a growing child with a C5 spinal cord injury. Spinal Cord **39**(2), 118–123 (2001)
65. Snoek, G.J., Ijzerman, M.J., Groen, F., et al.: Use of the NESS Handmaster to restore hand-function in tetraplegia: clinical experiences in ten patients. Spinal Cord **38**, 244–249 (2000). https://doi.org/10.1038/sj.sc.3100980
66. Son, B.C., Kim, D.-R., Kim, Y., Hong, J.T.: Phrenic Nerve Stimulation for Diaphragm Pacing in a Quadriplegic Patient. J Korean Neurosurg Soc. **54**(4), 359–362 (2013)

67. Stein, J., Narendran, K., McBean, J., et al.: Electromyography-controlled exoskeletal upper-limb-powered orthosis for exercise training after stroke. Am. J. Phys. Med. Rehabil. **86**, 255–261 (2007)
68. Stein, R.B., Everaert, D.G., Thompson, A.K., Chong, S.L., Whittaker, M., et al.: Long-term therapeutic and orthotic effects of a foot drop stimulator on walking performance in progressive and nonprogressive neurological disorders. Neurorehabil. Neural Repair **24**(2), 152–167 (2010). https://doi.org/10.1177/1545968309347681
69. Stroke Statistics: On-line reference (2018). http://www.strokecenter.org/patients/about-stroke/stroke-statistics/. Accessed on 11 April 2018
70. Sukhvinder, K.-R., Verrier, M.: A synthesis of best evidence for the restoration of upper-extremity function in people with tetraplegia. Physiother. Can. **63**(4), 474–489 (2011)
71. Taylor, P.N., Wilkinson Hart, I.A., Khan, M.S., et al.: Correction of footdrop due to multiple sclerosis using the STIMuSTEP implanted dropped foot stimulator. Int. J. MS Care **18**, 239–247 (2016)
72. Taylor, P., Barrett, C., Mann, G., et al.: A feasibility study to investigate the effect of functional electrical stimulation and physiotherapy exercise on the quality of gait of people with multiple sclerosis. Neuromodulation **17**(1), 75–84 (2014)
73. Taylor, P., Esnouf, J., Hobby, J.: Pattern of use and user satisfaction of neuro control freehand system. Spinal Cord **39**, 156–160 (2001). https://doi.org/10.1038/sj.sc.3101126
74. Taylor, P.N., Burridge, J.H., Dunkerley, A.L., Wood, D.E., Norton, J.A., et al.: Clinical use of the Odstock dropped foot stimulator: its effect on the speed and effort of walking. Arch. Phys. Med. Rehabil. **80**(12), 157783 (1999)
75. Tu, X.K., Zhou, X., Li, J.X., Su, C., Sun, X.T., et al.: Iterative learning control applied to a hybrid rehabilitation exoskeleton system powered by PAM and FES. Cluster Comput. J. Networks Softw. Tools Appl, **20**(4), 2855–2868 (2017)
76. Turk, R., Burridge, J., Davis, R., Cosendai, G., Sparrow, O.: Therapeutic effectiveness of electric stimulation of the upper-limb poststroke using implanted microstimulators. Arch. Phys. Med. Rehabil. **89**, 1913–1922 (2008). https://doi.org/10.1016/j.apmr.2008.01.030
77. Van den Brand, R., Heutschi, J., Barraud, Q., DiGiovanna, J., Bartholdi, K., et al.: Restoring voluntary control of locomotion after paralyzing spinal cord injury. Science **336**(6085), 1182–1185 (2012)
78. Wood, D.E., Harper, V.J., Barr, F.M.D., Taylor, P.N., Phillips, G.F., et al.: Experience in using knee angles as part of a closed-loop algorithm to control FES-assisted paraplegic standing. In: Proceedings of 6th International Workshop on FES: Basics, Technology and Application, Vienna, Austria, pp. 137–140 (1998)
79. Xu, R., Jiang, N., Mrachacz-Kersting, N., et al.: A closed-loop brain-computer interface triggering an active anklefoot orthosis for inducing cortical neural plasticity. IEEE Trans. Biomed. Eng. **61**, 2092–2101 (2014)
80. Young, B.M., Nigogosyan, Z., Nair, V.A., Walton, L.M., Song, J., Tyler, M.E., Edwards, D.F., Caldera, K., Sattin, J.A., Williams, J.C., Prabhakaran, V.: Case report: post-stroke interventional BCI rehabilitation in an individual with preexisting sensorineural disability. Front. Neuroeng. **7**, 18 (2014). https://doi.org/10.3389/fneng.2014.00018

Additional Reading Section (Resource List)

81. Chang, S.N., Nijholt, A., Lotte, F. (eds.): Brain-Computer Interfaces Handbook: Technological and Theoretical Advances. Boca Raton, CRC Press (2018). ISBN 978-1-498-77343-0
82. Chapin, J.K., Moxon, K.A. (eds.): Neural Prostheses for Restoration of Sensory and Motor Function. CRC Press, Boca Raton (2001). ISBN 0-8493-2225-1
83. Diez, P. (ed.): Smart Wheelchairs and Brain-Computer Interfaces. Academic Press, London (2018). ISBN 978-0-12-812892-3

84. De Horch, K.W., Dhillon, G.S. (eds.): Neuroprosthetics: Theory and Practice. Series on Bioengineering & Biomedical Engineering, vol. 2. World Scientific Publishing Co. Pte. Ltd., Singapore (2004). ISBN 981-238-022-1
85. DiLorenzo, D.J., Bronzino, J.D. (eds.): Neuroengineering. CRC Press, Boca Raton (2008). ISBN 978-0-8493-8174-4
86. Fazel, R. (ed.): Recent Advances in Brain-Computer Interface Systems. IntechOpen (2011). ISBN 978–953-307-175-6
87. Finn, W.E., LoPresti, P.G. (eds.): Handbook of Neuroprosthetic Methods. CRC Press, Boca Raton (2003). ISBN 0-8493-1100-4
88. Freeman, C.: Control System Design for Electrical Stimulation in Upper Limb Rehabilitation: Modelling, Identification and Robust Performance. Springer, Heidelberg (2016)
89. Garcia, B.M. (ed.): Motor Imagery: Emerging Practices, Role in Physical Therapy and Clinical Implications. Nova Science Publication, Inc., New York (2015). ISBN 978-1-63483-163-5
90. Graimann, B., Brendan, A., Pfurtscheller, G. (eds.): Brain-Computer Interfaces: Revolutionizing Human-Computer Interacation (The Frontiers Collection). Springer, Berlin (2010). ISBN 978-3-642-02090-2
91. Kilgore, K. (ed.): Implantable Neuroprostheses for Restoring Function. Elsevier Ltd., Amsterdam, Boston (2015). ISBN 978-1-78242-101-6
92. Kralj, A., Bajd, T.: Functional Electrical Stimulation: Standing and walking After Spinal Cord Injury. CRC Press, Boca Raton (1989). ISBN 0-8493-4529-4
93. Krames, E.S., Peckham, P.H., Rezai, A.R. (eds.): Neuromodulation: Comprehensive Textbook of Principles, Technologies, and Therapies, vol. 1, 2nd edn. Elsevier Ltd., London (2018). ISBN 978-0-12-802766-0
94. Levi, T., Bonifazi, P., Massobrio, P., Chiappalone, M. (Eds): Closed-loop systems for next generation neuroprostheses. Frontiers (2018). ISBN 978-2-88945-466-2
95. Malmivuo, J., Plonsey, R.: Bioelectromagnetism: Principles and Applications of Bioelectric and Biomagnetic Fields. Oxford University Press, New York (1995)
96. Niedermeyer, E., da Silva, F.L., (eds.): Electroencephalography: Basic Principles, Clinical Applications, and Related Fields. Lippincott Williams & Wilkins (2005). ISBN 978-0-7817-5126-1
97. Phillips, C.A.: Functional Electrical Rehabilitation: Technological Restoration after Spinal Cord Injury. Springer, Berlin (1991). ISBN 978-1-4612-7796-5
98. Popovic, D., Sinkjaer, T.: Control of Movement for the Physically Disabled. Springer, Berlin (2000). ISBN 978-1-4471-1141-2
99. Sandrini, G., Homberg, V., Saltuari, L., Smania, N., Pedrocchi, A.: Advanced Technologies for the Rehabilitation of Gait and Balance Disorders. Biosystems & Biorobotics. Springer International Publishing AG, Berlin (2018). ISBN 978-3-319-72735-6
100. Schalk, G., Mellinger, J. (eds.): A Practical Guide to Brain-Computer Interfacing with BCI2000. Springer, London (2010). ISBN 978-1-84996-091-5
101. Vuckovic, A., Pineda, J., Lamarca, K., Gupta, D., Guger, C. (eds.): Interaction of BCI with the Underlying Neurological Conditions in Patients: Pros and Cons. Frontiers Media SA, Lausanne (2015). ISBN 978-2-88919-489-6
102. Wolpaw, J.R., Wolpaw, E.W. (eds.): Brain-Computer Interfaces: Principles and Practice. Oxford University Press, Oxford (2012). ISBN 978-0-195-38885-5
103. Yu, W., Chattopadhyay, S., Lim, T.-C., Acharya, U.R.: Advances in Therapeutic Engineering. CRC Press, Boca Raton (2013). ISBN 978-1-4398-7174-4

Chapter 4
Mathematical Models Used in Intelligent Assistive Technologies: Response Surface Methodology in Software Tools Optimization for Medical Rehabilitation

Oana Geman, Octavian Postolache and Iuliana Chiuchisan

Abstract Assistive Technology (AT) refers to any commercially, modified, or customized item, equipment, or system that is used to enhance, maintain or improve the functional capabilities of people with disabilities. Assistive technologies should be viewed as a series of products and services through which people with different types of disabilities achieve their own goals by facilitating an independent life. The scope of support technologies extends to those products and services for three categories of people: people with disabilities, the elderly and people with chronic illnesses. Ambient Assistance Living (AAL) is a subcategory of environmental intelligence, which refers to the use of intelligent environmental techniques, processes and technologies to enable the elderly to live independently for as long as possible without intrusive behaviors. The Exergaming Platform presented in this chapter is a recovery application, designed for upper limb rehabilitation, that helps patients with locomotor disabilities and not only, transforming the unpleasant physical therapy into a fun game. The platform transforms the traditional games into video game-based exercises and drives patients to exercise correctly, while monitoring them. A RSM tool for Medical Rehabilitation based on an Exergaming Platform that uses a Microsoft Kinect Sensors, is presented in this chapter. The development of this application was made in cooperation between a research team from Institute de Telecommunication from Lisbon, Portugal and a research team from the University of Suceava, Romania. In order to find the score for a subject with a locomotor system leisure or neurodegenerative disorders, we used RSM Methodology and we optimized the exergame using statistical and mathematical. During the model developing, the data analysis

O. Geman (✉)
Department of Health and Human Development,
Stefan cel Mare University Suceava, Suceava, Romania
e-mail: oana.geman@usm.ro

O. Postolache
ISCTE-Instituto Universitário de Lisboa and Instituto de Telecomunicações,
Lisbon, Portugal

I. Chiuchisan
Computers, Electronics and Automation Department,
Stefan cel Mare University Suceava, Suceava, Romania

H. Costin et al. (eds.), *Recent Advances in Intelligent Assistive Technologies: Paradigms and Applications*, Intelligent Systems Reference Library 170,
https://doi.org/10.1007/978-3-030-30817-9_4

has shown that the RSM can be a good candidate for optimization the application. The application has demonstrated that the Response Surface Methodology (RSM) is a useful instrument in the prediction of the patients variable scores.

Keywords Assistive technologies · Mathematical models · Response surface method · Exergaming · Rehabilitation

4.1 Introduction

According to the World Health Organization [48], "about one billion people today need Intelligent Assistive Technologies (IAT) to lead a complete life, either because they are elderly or because they have a disability". The number of aging people worldwide is rising faster than other age groups. According to forecasts, in 2050 the number of elderly people in the world will exceed the number of children (Fig. 4.1).

It is estimated that by 2050 the number of assistive devices will increase to two billion. The World Health Organization [48] estimates that only one in 10 people in need of these vital technologies have access to them due to lack of availability and awareness, but also because of high costs. The solution to the problem of demographic aging is to keep older people active in their communities. To help the elderly to remain active, we need to give them the chance to participate fully in society, to create employment opportunities, to facilitate their access to volunteer activities (caring

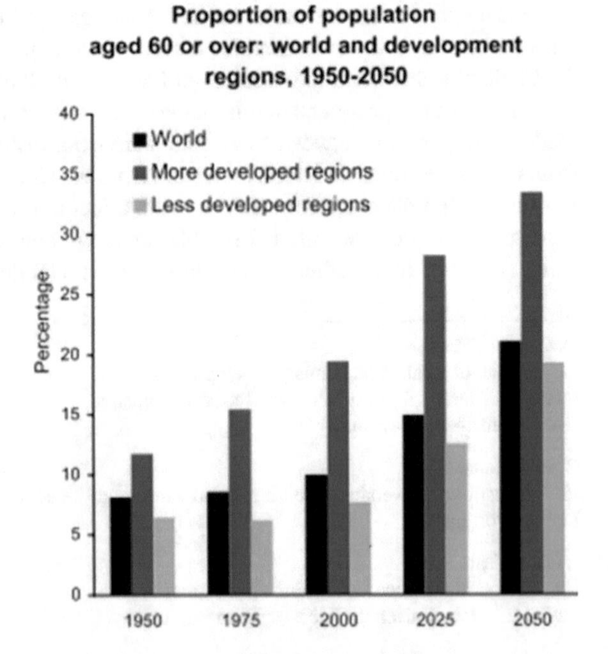

Fig. 4.1 Population aged 60 or over. *Source* European Commission—Active ageing, 2017

for family members), to create conditions for them to lead an autonomous life by adapting housing, means of transport, infrastructure, etc.

4.2 State-of-the-Art in Intelligent Assistive Technologies

The Ambient Intelligence (AmI) goal is to support "humans in achieving their everyday objectives by enriching physical environments with networks of distributed devices, such as sensors, wearable devices, and computational resources". AmI is not only "the convergence of various technologies (i.e. sensor networks and industrial electronics) and related research fields (i.e. pervasive, distributed computing, and artificial intelligence), but it represents a major effort to integrate them and to make them really useful for everyday human life" [33]. AAL includes technological assistance systems, promising support for an independent life in old age and an increased quality of life [6–10]. AAL is involved in the development of smart home, mobile devices, wearable sensors, smart fabrics, and assistive robotics (Fig. 4.2). Wearable technology, sensors, apps, and online services can be a useful tool for elderly people to complete the everyday tasks. The systems included in AAL environments can adjust automatically, proactively and situation-specific to the user's needs and goals. In the future, AAL systems will be expanded to include specially developed care robots [2].

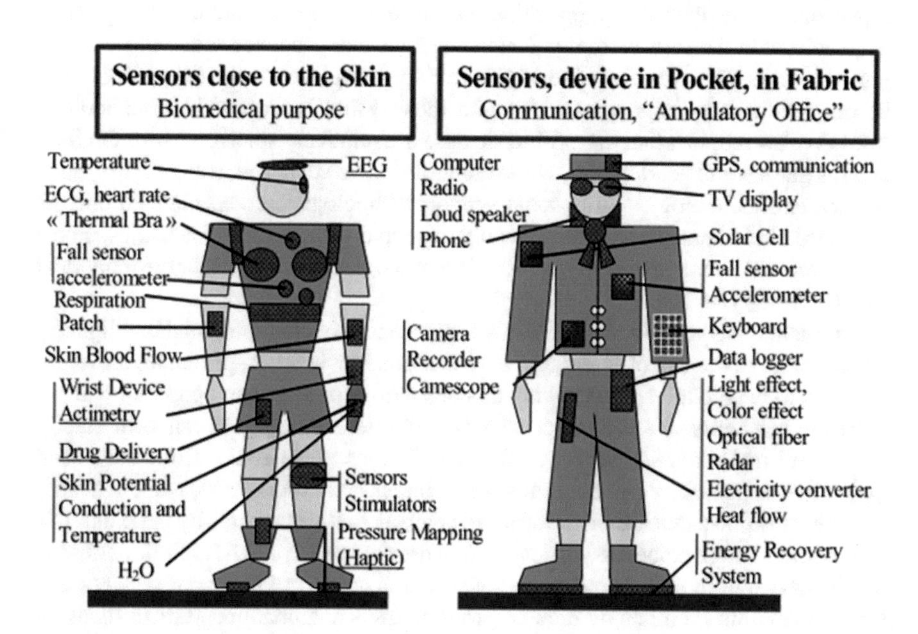

Fig. 4.2 Wearable devices [5]

Assistive technology (AT) includes assistance, adaptation and rehabilitation for the elderly or disabled, but also includes the process of selecting, using and locating them. People with disabilities most often encounter difficulties in performing day-to-day activities independently or with assistance. AT refers to any commercially, modified, or customized item, equipment, or system that is used to enhance, maintain or improve the functional capabilities of people with disabilities. Assistive Technology promotes the independence of people with disabilities, enabling people to perform tasks that they have not been able to accomplish or they are very difficult, providing improvements or methods of interaction with the technology that will carry out the tasks. AT can offer support to the persons with disabilities, elderly, people with diabetes and stroke or people with mental illness, including dementia and autism, in several categories of deficiencies: (1) Mobility deficiencies; (2) Visual impairments; (3) Hearing impaired; (4) Cognitive impairments.

Assistive technology can have a positive impact on the health and well-being of a person and his/her family. The World Health Organization [48] coordinates Global Assistance Technology Assistance (GATE), which exists to improve access to high-quality, accessible access technology for everyone everywhere. The GATE initiative will strengthen the overall health strategy of the world health organization on people centered and integrated health services as well as action plans on non-communicable diseases, aging and health, disability and mental health.

Assistive technologies include tools used to enhance the independent functioning of people with physical limitations or cognitive dysfunction. This access technology, represents all the mechanical, optical, electronic and computer solutions that provide to people with sensory or motor deficiencies an independence, allowing them to perform tasks that are otherwise impossible to perform or with major difficulties accomplishment, by improving or changing the way they interact with the technology needed to accomplish that task. Some devices are relatively small and very familiar, such as glasses and hearing aids, while others are more advanced, using cutting-edge science and technology. While some support technologies are specialized devices (standard market products) dedicated only to people with specific disabilities, others have more common features to traditional technologies, configured for the individual user (Fig. 4.3) [30, 32].

Ambient Assistance Living (AAL) is a subcategory of environmental intelligence, which refers to the use of intelligent environmental techniques, processes and technologies to enable the elderly to live independently for as long as possible without intrusive behaviors. AAL focuses on supporting technology that can help elderly to improve mobility and everyday life tasks. The results of AAL efforts have been innovative, especially when data show a real increase in elderly care, while reducing staffing workflows that are often intrusive (such as routine controls by patients). IoT technology is often embedded in devices that are not intrusive. AAL solutions always play a vital role in facilitating operational optimization of healthcare services and workflow within facilities by reducing routine tasks and enabling staff to focus on priority services.

Fig. 4.3 Examples of exergames that use Microsoft Kinect™ Sensor [30, 32]

4.2.1 Rehabilitation Using Exergaming

Given the recent development in ICT, an increasing number of wearable devices have been developed and connected in order to improve healthcare delivery. Healthcare providers are connected through mobile computers, tablets, smartphones, and other portable devices. The researchers in fields such as ICT and Internet of Things (IoT) are now working on "developing further connectivity to improve not only communications between healthcare givers and patients but also the real-time monitoring of patient's health" [44]–[47].

The Internet of Things (IoT) is "a concept that aims to define a world in which all objects (cars, home appliances, lighting systems, mobile devices, portable devices, smartphones etc.) are connected to each other via the Internet. Every object, even the human body, can become a small part of the IoT if it is endowed with certain electronic components" [14]. These parts may vary, depending on the purpose of the object, but fall into two broad categories:

(1) The object must be capable of data acquiring, usually through the sensors;
(2) The object must be capable of transmitting this data via the Internet.

Together with other inventions such as "Cloud computing, smart grids, nanotechnology and robotics, the Internet of Things world provides a huge step forward towards an economy characterized by increased efficiency, productivity, safety and profit" [37].

The developments in Information System (IS) promise "an improvement of clinical and administrative reporting capabilities, operational efficiency, communication among health professionals, communication with patients, data accuracy and effectiveness of physiotherapy based on a serious game and augmented reality" [35]. Medical recovery through game play or exergaming consists of a series of computer games that help the patient to test his limits, e.g. to catch, guide and position different objects on the screen.

Professor Octavian Postolache and his team of researchers from Institute of Telecommunication from Lisbon, Portugal [34–43] in collaboration with a research team from the University of Suceava, Romania [3–5, 11–18] have developed and tested an exergaming platform for medical rehabilitation. The Exergaming Platform presented in this chapter is a recovery application that helps patients, such as people with locomotor disabilities, elderly, people with neurodegenerative disorders and not only, transforming the unpleasant physical therapy into a fun game. The platform transforms the traditional computer games into video game exercises and drives patients to exercise correctly, while monitoring them. The game was designed and implemented for upper limb rehabilitation. The user's avatar pass through an orchard with apples, located in different position, and has to picks up red or green apples according to their level. Different angles of inclination of the neck, hand, shoulder, etc. are measured during the game and a final score is generated [11] (Fig. 4.4).

The result is a faster and more interactive recovery therapy method for patients. To interact with patients, the exergaming platform uses Microsoft Kinect, a camera that can track users' movements remotely without requiring additional sensors attached to the human body. The platform is easy to use and offers video games created according to the needs and medical problems of each patient. The patient needs a computer and a Microsoft Kinect video camera.

Xbox is a console that recognizes the body movements, which is different from other consoles that require a kind of physical controller to keep in hands. While

Fig. 4.4 The "Apple Harvesting" 3D Game user interface including the Avatar [34]

the patient is playing a particular game, the Microsoft Kinect camera monitors all the movements and can give the doctor important information about his progress. Microsoft Kinect™ Sensor is characterized by specific hardware and firmware that assure the generation of three-dimensional images of objects and may recognize the person moving in between those objects. About 200 moves are pre-programmed, so the software can predict the body movement [21–29]. Recognition of body and limb movement is based on the analysis of almost 20 points associated with the body's joints. In addition to cameras, motion detection, and recognition software, Microsoft Kinect includes four microphones that analyze the environment sound. Together with dedicated speech recognition software, the user may control the platform by voice.

This recovery application measures certain parameters such as the movement speed, angles of inclination of the neck, hand, shoulder etc. The physiotherapist has access to all the measurements that can show how effective is the therapy for the patient. Based of this information, he can prescribe or adapt the exercises, depending on the patient's evolution, and through the feedback provided by the recovery application.

This exergaming platform presented in this chapter can be used also in physically rehabilitation in case of patients that sufferd from a vascular accident. This innovative method of physically recovering, in case of the patients affected by a stroke, is based on the motion sensors used by the Xbox series electronic games. By developing video games tailored to the needs of a specific rehabilitation therapy, patients will be able to carry out the physical exercise required in a more engaging way.

Recovery of limb mobility after a stroke is possible but requires many months of daily exercise (preferably 45–60 min) under medical supervision. This has a number of drawbacks, such as the need to move patients to recovery sessions, the lack of the necessary number of specialized medical staff, and so on. For these reasons, around 80% of patients will never regain hand and arm mobility, which drastically limits their independence and their ability to perform the basic daily needs.

Recovery of limb mobility after a stroke is possible but requires many months of daily exercise (preferably 45–60 min) under medical supervision. This has a number of drawbacks, such as the need to move patients to recovery sessions, the lack of the necessary number of specialized medical staff, and so on. For these reasons, around 80% of patients will never regain hand and arm mobility, which drastically limits their independence, their ability to cope with basic daily needs, in short, greatly reduces their level of living. Patients who will properly physically recover during the first few months after the stroke can be rehabilitated almost completely, gaining their mobility, being able to maintain themselves and even returning to work. The platform allows the physician to remotely monitor the patient's progress and adapt the type and level of exercise difficulty to individual needs. The exergames are designed to ensure the training of both the muscles involved in ample movements and those for fine movements.

4.3 Mathematical Models: Response Surface Methodology—Algorithm Implementation and Algorithm Case Study: Performance Evaluation of Exergame-Based Rehabilitation

4.3.1 General Concepts on Correlation and Statistical Regression

Biostatistics is a branch of Mathematical Statistics applied to the study of biological processes, establishing laws of behavior based on the analysis of a limited number of data, obtained by specific selection and random theory, or by analogous experiments associated with those processes.

The statistical analysis is based on certain approximation methods specific to the theory of estimation and on the verification of hypotheses formulated in accordance with the concrete requirements of the statistical survey. In essence, all properties of statistical approximation represent the application of two fundamental results from the Probability Theory, the law of large numbers and the central limit theorem. In practice, we often need to analyze large collections of experimental results or other quantities, and we can not always examine all amount of data. In such cases, instead of examining the entire set of data (called population), we may extract some conclusions after examining only a part of data, taken at random (selection). The procedure is called full selection, and the process of extrapolating the findings to the entire population is known as statistical inference.

4.3.2 Linear Correlation and Regression Analysis Based on Ungrouped Data

Suppose that, as a result of angle measurement performed using Microsoft Kinect™, we have n pairs of observed values (x_i, y_i) corresponding to the factors (indicators) x and y, respectively, of the statistical study of a general collectivity, on which two-dimensional selection was made. On the basis of these data, we will determine the dependence of the two study factors by means of two probabilistic measures called *correlation* and *regression*. If we assume that the relationship of dependence is linear, then the theoretical analytical theorem in which y varies with respect to the values of x, is a function of first degree, whose graph will be a straight line, called the *theoretical line of regression* of the variable y in relation to the values of the variable x, where:

$$y = \alpha + \beta x \tag{4.1}$$

The slope β of this line is called the *correlation coefficient* and represents a measure of the intensity of the stochastic dependence of the two indicators. The simple linear regression analysis leads to the determination of the coefficients of the regression line a and b using the least squares method, and also it leads to the verification of the calculated estimates, using statistical tests to check the simple hypotheses regarding the coefficients [31].

4.3.3 Response Surface Methodology in Software Tools Optimization for Medical Rehabilitation

The statistical programming of the experiments, respectively the analysis of the influence of the process parameters on the final results, can be performed using the Surface Response Method. The use of this method allows a hierarchy of the considered factors influence and the highlighting of the interaction between the independent variables [49].

The experimental research of a process is to highlight the most significant parameters that can influence the process under consideration, and to establish the links between them and the answers they have obtained. The response surface method "considers the connection between the process parameters, and their effects on the studied phenomenon as surfaces in the three-dimensional space of the variables, called response surfaces. This assumes that for each value of the considered parameters, a value for the dependent function will be determined, which will be on the response surface" [20]. The problems posed by determining a regression function are mainly the following:

(a) statistical programming of the experiments;
(b) establishing the function (model);
(c) calculation of regression coefficients;
(d) regression analysis;
(e) determination of statistical errors and confidence intervals for the dependent variable.

The fundamental objective of the research is to model the investigated system. The mathematical and analytical modeling of the studied system involves determining the action of the influence factors x_1, x_2, x_3, …, x_k, …, x_f on the objective function (where $k = 1, 2, 3, \ldots, f$, and f is the number of factors). Explaining the form of the analytical function requires the indication of the functional dependence:

$$y = f(x_1, x_2, x_3, \ldots, x_k, \ldots, x_f,) \tag{4.2}$$

However, in the case of complex technical systems, the mathematical modeling is extremely difficult, which is why in such situations the experimental modeling or mixed analytical-experimental modeling is used.

Experimental modeling of the processes or investigated systems involves defining the objective of the research, choosing the objective function, identifying the influence factors, checking the suitability of the model, etc. A basic requirement imposed on models, both analytical and experimental, is that they have the ability to reflect the most properly the investigated system. This requires that the values of the objective function evaluated using the model (the "replica" of the investigated real system) represent the most accurate image of the technical system's values (object, phenomenon or process). Simultaneously, experimental modeling should provide experimental direct information on the investigated system to reach the optimal field of objective function. In the experimental research strategy we have observed four consecutive stages, which are iteratively included in complete investigation cycles to achieve the objective of experimental research—determination of the optimal experimental model—to a minimum number of experiments performed without reducing its estimation accuracy. The basic stages are:

1. Adoption of the a priori mathematical model for the studied system, which establishes the functional link between the objective function and the influence factors;
2. Conceiving the experimental program, in accordance with the number and level of influence factors, as well as the number of planned replies;
3. Effective realization of the research program dedicated to the experimental cycle;
4. Processing the experimental results by evaluating the statistical data of the observation data, testing the suitability of the model and, if necessary, correcting the model to define the optimal model of the investigated system.

Adopting an optimal strategy for carrying out the experiments modeling involves designing an implementation program and an optimal experimental research plan. The experimental research strategy conducted according to the classic *Gauss-Seidel* model follows the "one factor at a time" algorithm (a "unifactorial" research). This means that at one point the value of a single influence factor is changed and constant values are given for the other factors, by which the volume of experimentation increases significantly. The *Box-Wilson* model factor experimentation strategy follows the "all factors in every moment" algorithm. This means that in factorial experiments for each experimental test, the value of all influence factors is changed, which can greatly reduce the volume of experimentation.

This modern strategy has the following main features:

(a) The information is acquired progressively through the experience, requiring a minimum number of experiences to formulate the conclusions;
(b) The method allows to provide information about the direction of movement of determinations, for obtaining the optimal range of the objective function.
(c) The maximum estimation accuracy of the model is obtained.

The relationship between the objective function y and the influence factors x_1, x_2, x_3, ..., x_k, ..., x_f can be described in the general form as:

$$y = \varphi(x_1, x_2, x_3, \ldots, x_k, z_1, z_2, z_3, \ldots, z_m, \beta_1, \beta_2, \beta_3, \ldots, \beta_d) \quad (4.3)$$

in which:

- $x_1, x_2, x_3, \ldots, x_k$ represent the natural factors of controllable influence;
- $z_1, z_2, z_3, \ldots, z_m$ are uncontrollable factors of influence, random stochastic, and generating random errors;
- $\beta_1, \beta_2, \beta_3, \ldots, \beta_d$ represent the statistical parameters, usually unknown, called coefficients of influence or regression.

In real situations, the shape and structure of the real model of the systems, in particular of the technological systems, are unknown. In the process of modeling the real system, the real values of the regression coefficients $\beta_1, \beta_2, \beta_3, \ldots, \beta_d$ for the given model with their statistical estimates $b_1, b_2, b_3, \ldots, b_d$ determined by the processing of the experimental data are replaced, and the influence of the random factors is included in the experimental error, aiming the minimization.

In this way, we move from the real model to the empirical (experimental) model in which the real response function is replaced by its statistical estimation, expressed by a function of the controllable factors and the statistical estimations of the real regression coefficients:

$$y = \varphi(x_1, x_2, x_3, \ldots, x_k, z_1, z_2, z_3, \ldots, z_m, b_1, b_2, b_3, \ldots, b_d) \tag{4.4}$$

In solving the problem of experimental modeling, we try to express the connection between the objective function and the influence factors in the polynomial form. In general, a polynomial model of first order is expressed by the relation of the form:

$$y = b_0 + \sum_{i=1}^{k} b_i \cdot x_i \tag{4.5}$$

$$y = b_0 + \sum_{i=1}^{k} b_i \cdot x_i + \sum_{i=1}^{k} b_{ii} \cdot x_i^2 + \sum_{i,j=1, i \neq j}^{k} b_{ij} \cdot x_i \cdot x_j \tag{4.6}$$

RSM is "useful for programs modeling and analysis in which a response of interest is influenced by several variables and the objective is to optimize this response. In our exergaming platform we needed to find the levels of spine degree (x_1) and neck degree (x_2) to maximize the score value (y) of the exergame" [11].

$$y = f(x_1, x_2) + \varepsilon \tag{4.7}$$

The geometric representation of the functional dependence, called the response surface, gives the estimation of the real surface. Representation of this function is possible only for the case of the dependence of y by no more than two influence factors. In general situations, the two-dimensional sections of the response hyperspace can be viewed. The response surface is:

$$\eta = E(y) = f(x_1, x_2) \tag{4.8}$$

Performing the optimal factorial experiment has the purpose to define the data necessary to calculate the effects caused by the influence factors on the objective function, as well as the intensity of the interactions between the influence factors. Effects and interactions are expressed using the coefficients of the adopted model. The number of coefficients determines the volume of experimentation, so the minimum degree of polynomial is chosen according to the number of unknown coefficients.

In case of an objective function y dependent on two factors of influence x_1 and x_2, the experimental model can be represented by an first order polynomial function:

$$y = b_0 + b_1 \cdot x_1 + b_2 \cdot x_2 \tag{4.9}$$

and also usable in areas where the response surface can be approximated to a flat surface, or a second order polynomial function, usable in extreme areas where the curvature of the response surface is accentuated:

$$y = b_0 + b_1 \cdot x_1 + b_2 \cdot x_2 + b_{12} \cdot x_1 \cdot x_2 + b_{11} \cdot x_1^2 + b_{22} \cdot x_2^2 \tag{4.10}$$

The function f is unknown and approximate the true relationship between y and the independent variables by the lower-order polynomial model.

$$y = \beta_0 + \beta_1 x_1 + \Lambda + \beta_k x_k + \varepsilon \tag{4.11}$$

$$y = \beta_0 + \sum_{i=1}^{k} \beta_i x_i + \sum_{i=1}^{k} \beta_{ii} x_i^2 + \sum_{i<j}^{k} \beta_{ij} x_i x_j + \varepsilon \tag{4.12}$$

In first-order factorial experiments ("screening experiments"), the determination of coefficients of the first degree polynomial function involves selecting the influence factors for two levels of variation, for the minimum and maximum levels, requiring a $n = 2\,k$ volume of experiments. The optimal strategy provided by process involves the use of a minimal number of attempts made by an optimal strategy and the regression polynomial function corresponding to this type of experiment provides information on the direction to be optimized.

An important stage in modeling based on experiments is the verification of the suitability of the estimated polynomial model, i.e. the consistency between the results of the measurements and those estimated using the model. The number of regression coefficients that can be calculated is equal to the volume n of the experiment. Regardless of the fact that we use the complete factorial experiments or fractional factorial experiments (truncated), the influence factors are always assigned only two levels of variation (minimum and maximum) to determine the regression coefficients (including b_0).

For the continuous objective function we choose the dimensions of the experimentally explored subdomain, that is, the areas of variation of the influence factors, which can be determined on the basis of the a priori information. Starting from this area based on the information obtained, the experiment continues sequentially following the direction of the maximum slope (gradient direction) on the response surface, moving to other appropriately sized sub-domains, until the optimal domain containing the point of interest (which is usually an extreme point).

There are situations when the response surface has a too high curve, for the modeling to produce the satisfactory results in terms of precision estimation. In these situations, is turn to higher order polynomials model, preferably of the second order, until the optimum is identified. These models, called higher order models, can be explained either by directing influence factors on three levels of variation (which greatly increases the volume of the experiment by $n = 3\ k$ and complicates the processing of its results), or by resorting to the so-called central-composed experiments. Based on the experimental points, it can find a simple mathematical expression of a surface in space with $n + 1$ dimensions so that the surface approximates optimally, after a certain criterion, the set of experimental points. The expression deduced, called the regression function, will not coincide with the theoretical one, but it will approximate it sufficiently precisely to allow its use in practical applications or even as the initial hypothesis in some theoretical studies.

4.3.4 Case Study: Exergaming Platform

In order to find if the patient has progressed in physical rehabilitation therapy using the exergaming platform presented in Sect. 2.1, the final score should be increased. In order to find the score that a patient with a locomotor system disorder or stroke can achieve, we have optimized the exergame by applying Response Surface Methodology (RSM) and Multiple Nonlinear Regression, using the *Expert Design Experimental* software program [11]. An optimization process was made in order to adjust a second-degree model in the experiment results because the approximation of the response area in the optimal (extremity) region through a first-degree model was unsatisfactory in our case (Fig. 4.5). We used the response surface method to optimize the initial conditions for improving the functional properties of the exergaming application [11].

4.3.4.1 Measurement System Analysis

The software programs used to implement the mathematical model in our system (exergaming application) are *DOE Expert Design Software Package*, *Minitab 18*, and *IBM SPSS* for descriptive statistics. Data acquired from the exergaming application contains the following variables: *sessions_ID, red_apple, gree_apple, timeamp, body_parts_ID, angle, leftelbow, leftsholder, rightelbow, rightshoulder, neck, spine and score*. In this chapter, the Response Surface Method was used to optimize the exergaming application for medical rehabilitation, depending on the influence of two variables, the *neck*, and *spine*, on several indicators: *angle, leftelbow, leftsholder, rightelbow, rightshoulder*.

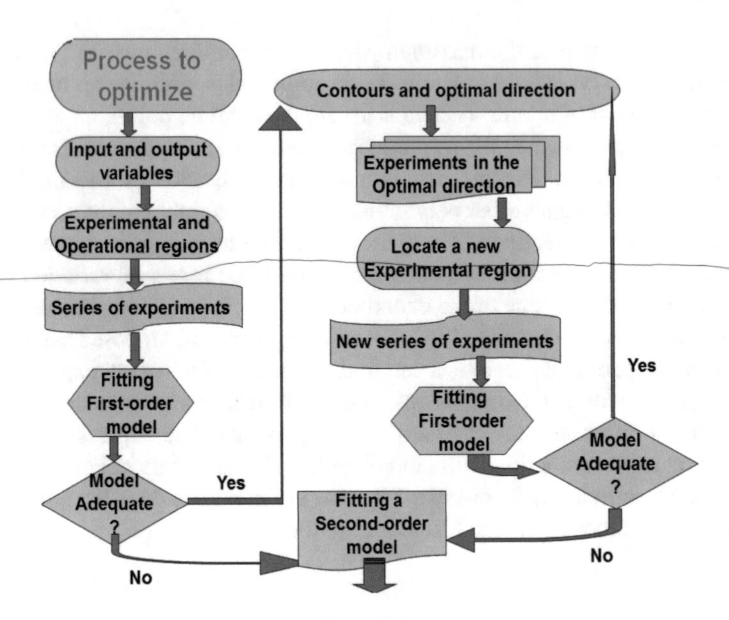

Fig. 4.5 Optimization process [31]

The experimental stages [19] of our study are:

1. experimental stage in order to extract data;
2. the modeling of the score obtained by healthy subjects (without neuro-muscular disorders);
3. the model prediction corrections that were evaluated using several statistical indicators, such as the determination coefficient and the square square error (RMSE);
4. extrapolate the RSM models that will be evaluated for new data sets.

In the diagram illustrated in Fig. 4.6, the steps of selecting the statistical method of the multidimensional series study, are presented. Their purpose is to identify and use possible links that can be manifested between two or more variables. We were interested by the existence of the links, the link intensity, the functional form of the link, its parameters on the researched experiment. The *Minitab program* provide the size of the sample used to create the statistics, and the number of observations from data that were missing.

Analyzing the link between the variables of the multidimensional distribution involves addressing the following problems, which can be classified in six stages to be covered in the statistical approach:

1. Organize the results of the population or sample observation in relation to the variables investigated;
2. Statistical analysis of the existence of the link;
3. Statistical analysis of the intensity of the link or degree of association between the observed variables;

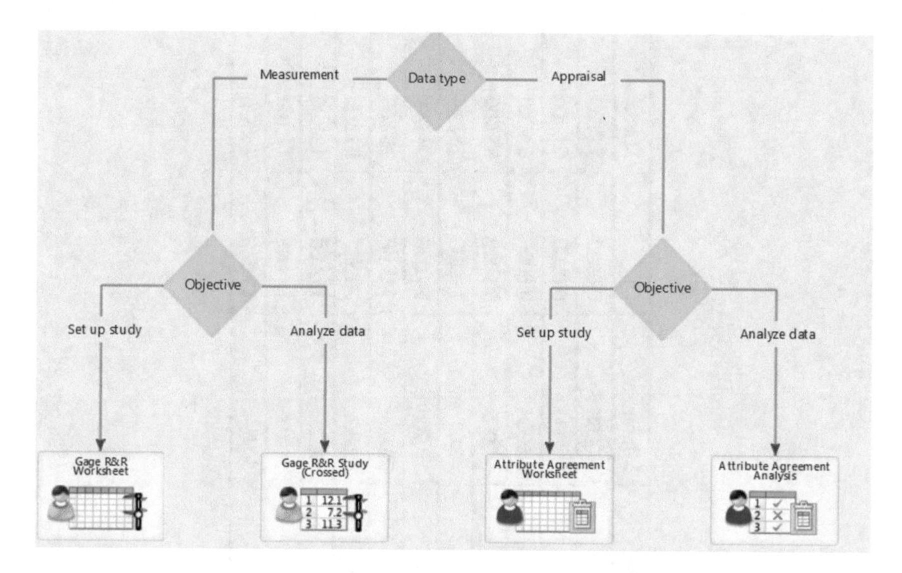

Fig. 4.6 Diagram of measurement system analysis (capture from Minitab 18 Software)

4. Formulate assumptions about the mathematical form of the link;
5. Estimating the regression function parameters;
6. Analyze the representativeness of the regression function.

These steps can be fully or partially covered, depending on the nature of the variables. In Tables 4.1 and 4.2, dataset variables for men and women, are presented.

Centrality, dispersion, location, symmetry indicators were calculated by using predefined functions and specific formulas using the *Descriptive Statistics option in Data Analysis Tools* (Fig. 4.7). The frequencies for angles are shown in Fig. 4.8.

4.3.4.2 Regression Analysis

The multiple regression function obtained by the minimum squares method is illustrated in Figs. 4.9, 4.10 and 4.11.

4.3.4.3 Experiments—Plan and Create

During the process of modeling the screening design we used the *Minitab software* that automatically randomizes the order of experimental data (Fig. 4.12). This randomization balances the effect of uncontrollable conditions and reduces the chance that these conditions will affect the results.

During the screening experiment it is necessary to perform the runs in random order. In order to complete the screening process we have to: complete all

Table 4.1 Statistics for men and women (dataset I)

Variable	Variance	CoefVar	Sum	Sum of Squares	Minimum	Ql	Median
Red_apple	0.2459	87.67	573.0000	573.0000	0.0000	0.0000	1.0000
Green_apple	0.2459	114.17	440.0000	440.0000	0.0000	0.0000	0.0000
Body_parts_ID	6.0981	30.48	8208.0000	72,678.0000	6.0000	6.0000	6.0000
Angle	209.004	18.28	80,105.000	6,545,975.000	55.000	70.000	85.000
Left elbow	520.18	15.14	152,606.00	23,516,146.00	47.00	139.00	157.00
Left shoulder	892.096	32.32	93,602.000	9,551,700.000	18.000	74.000	94.000
Right elbow	410.70	13.41	153,082.00	23,548,998.00	72.00	143.00	157.00
Right shoulder	1018.12	35.05	92,218.00	9,425,358.00	22.00	70.50	93.00
Neck	85.39	7.52	124,559.00	15,402,255.00	62.00	118.00	124.00
Spine	10.10	2.52	127,530.00	16,065,400.00	110.00	124.00	125.00
Score	3,801,395.1	66.68	2,961,900.0	125,073E+10	50.0	1350.0	2650.0

Table 4.2 Statistics for men and women (dataset II)

Variable	Q3	Maximum	Range	IQR	Mode	N for Mode	Skewness	Kurtosis	MSSD
Red_apple	1.0000	1.0000	1.0000	1.0000	1	573	−0.27	−1.93	0.2376
Green_apple	1.0000	1.0000	1.0000	1.0000	0	573	−0.27	−1.93	0.2376
Body_parts_ID	11.0000	11.0000	5.0000	5.0000	6	587	0.32	−1.90	6.4229
Angle	85.000	100.000	45.000	15.000	58	386	−0.18	−0.92	222.666
Left elbow	168.00	179.00	132.00	29.00	171	35	−1.32	1.61	355.96
Left shoulder	112.000	175.000	157.000	38.000	92	27	−0.24	−0.22	688.370
Right elbow	165.00	179.00	107.00	22.00	164	42	4.41	1.67	214.94
Right shoulder	114.00	178.00	156.00	43.50	95	19	−0.13	−0.42	765.94
Neck	129.00	157.00	95.00	11.00	128	66	−0.67	2.74	66.18
Spine	127.00	142.00	32.00	3.00	125	262	1.21	3.57	7.49
Score	4150.0	8950.0	8900.0	2800.0	250	14	0.66	−0.12	218,162.1

Descriptive Statistics

	N	Range	Minimum	Maximum	Mean		Std. Deviation	Variance	Skewness		Kurtosis	
	Statistic	Statistic	Statistic	Statistic	Statistic	Std. Error	Statistic	Statistic	Statistic	Std. Error	Statistic	Std. Error
angle	1063	5	6	11	8.41	.077	2.499	6.247	.074	.075	-1.998	.150
lelbow	1063	45	55	100	78.55	.450	14.661	214.943	-.224	.075	-.950	.150
lshoulder	1063	152	27	179	156.57	.649	21.170	448.150	-2.051	.075	5.114	.150
relbow	1063	141	18	159	95.29	.853	27.827	774.326	-.693	.075	.360	.150
rshoulder	1063	134	45	179	158.98	.540	17.613	310.225	-2.378	.075	8.453	.150
neck	1063	157	16	173	99.25	.848	27.653	764.661	-.620	.075	.024	.150
spine	1063	89	69	158	121.47	.331	10.800	116.648	-.266	.075	1.369	.150
score	1063	35	108	143	126.12	.126	4.114	16.925	.430	.075	2.179	.150
Valid N (listwise)	1063											

Fig. 4.7 Descriptive statistics (capture from Data Analysis Tools)

Statistics

		angle	lelbow	lshoulder	relbow	rshoulder	neck	spine
N	Valid	1013	1013	1013	1013	1013	1013	1013
	Missing	0	0	0	0	0	0	0
Mean		79.08	150.65	92.40	151.12	91.03	122.96	125.89
Std. Error of Mean		.454	.717	.938	.637	1.003	.290	.100
Median		85.00	157.00	94.00	157.00	93.00	124.00	125.00
Mode		85	171	92	164	95	128	125
Std. Deviation		14.457	22.807	29.868	20.266	31.908	9.241	3.177
Variance		209.004	520.179	892.096	410.705	1018.116	85.392	10.095
Skewness		-.184	-1.319	-.243	-1.408	-.134	-.671	1.213
Std. Error of Skewness		.077	.077	.077	.077	.077	.077	.077
Kurtosis		-.917	1.607	-.222	1.674	-.421	2.742	3.571
Std. Error of Kurtosis		.154	.154	.154	.154	.154	.154	.154
Range		45	132	157	107	156	95	32
Minimum		55	47	18	72	22	62	110
Maximum		100	179	175	179	178	157	142
Sum		80105	152606	93602	153082	92218	124559	127530

Fig. 4.8 Frequencies for angles (capture from Data Analysis Tools)

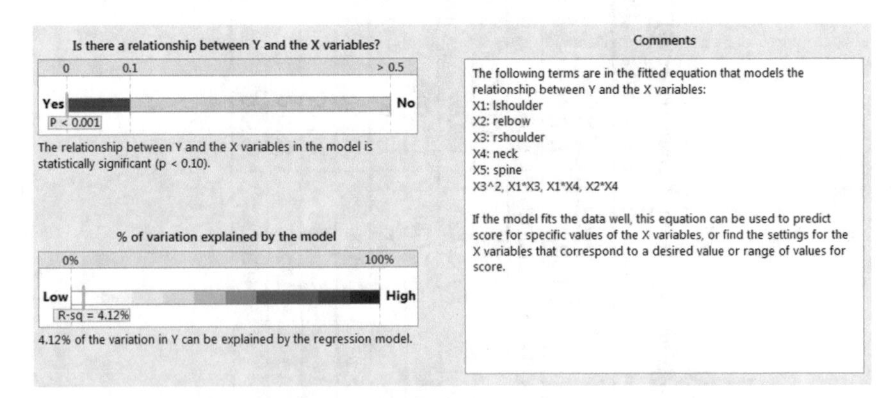

Fig. 4.9 Summary report for multiple regression

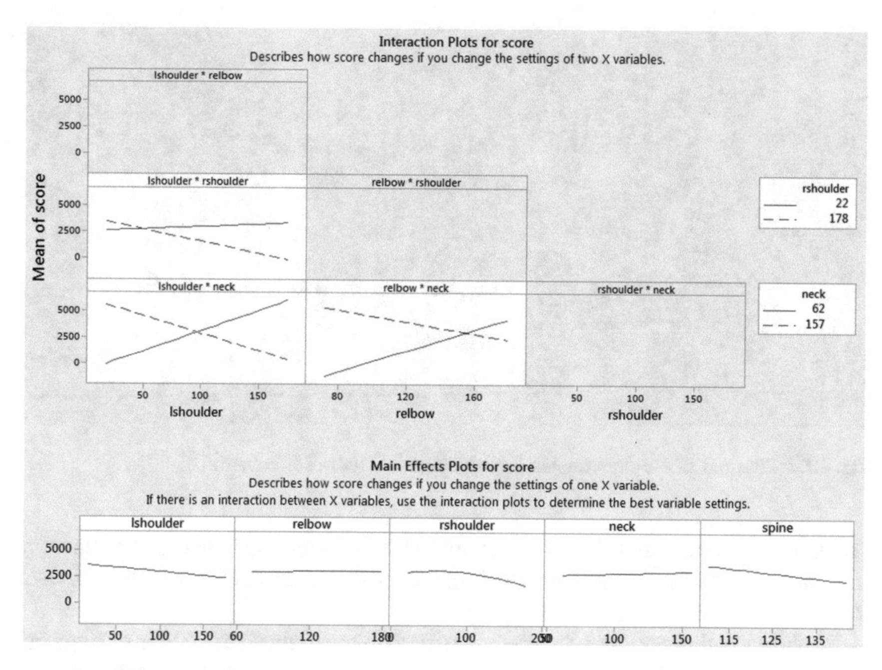

Fig. 4.10 Effect report for multiple regression

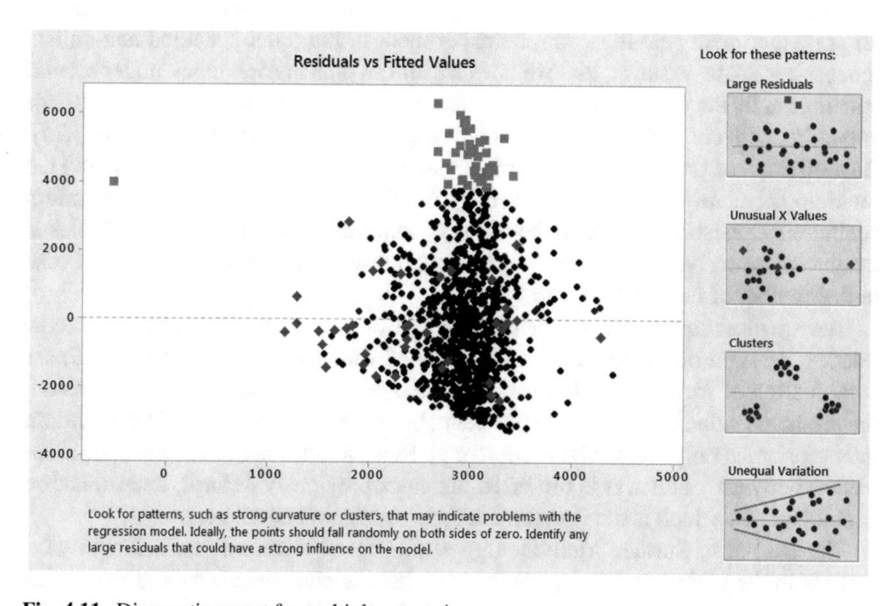

Fig. 4.11 Diagnostic report for multiple regression

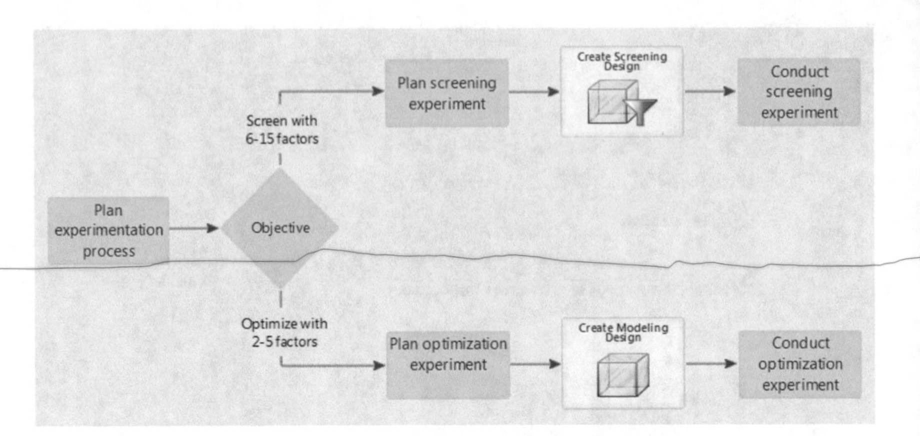

Fig. 4.12 Plan and create experiments (capture from Minitab 18 Software)

pre-experiment activities, run the experiment in the order specified and collect the response data, fit the screening model, and identify the critical few factors (five or fewer) to include in the modeling design.

Replicate and repeat measurements are both taken at the same factor settings but replicates are taken during separate experimental runs while repeats are taken during the same run. In order to complete the optimization process we have to: complete all pre-experiment activities, run the experiment in the order specified and collect the response data, assuring that we perform all the runs in each block under similar conditions, fit the linear model, and if the curvature is significant, add points for curvature, collect the response data and fit a quadratic model (Figs. 4.13 and 4.14). As a measure of the association between y and the set of variables x, is inserted the multiple correlation coefficient, noted by R, that can be defined as the maximum coefficient (Pearson) correlation between y and a linear combination of variables x. So, the calculated value of R is always positive and tends to increase with the increase in the number of independent variables.

The smallest square method can be thought of as a method maximizes the correlation between observed values and estimated values (these representing a linear combination of variables x). R-value close to 0 denotes an insignificant regression, the predicted values of regression not being better than those obtained by a random guess (or based only on the distribution of y). Since R tends to overestimate the association between y and x, it is preferred indicator previously defined, determination coefficient, R^2, which is the squared multiple correlation coefficient.

The Response Surface Methodology allows to approximate the behavior of a process in the vicinity of the optimum score but the challenge is to find the region within the range of the factors for which the RSM model is a good approximation and then locate the optimum score. During the model developing, the previous data analysis has shown that RSM can be a good candidate for optimization. A collection of mathematical and statistical techniques is used to develop a regression model (Figs. 4.15 and 4.16).

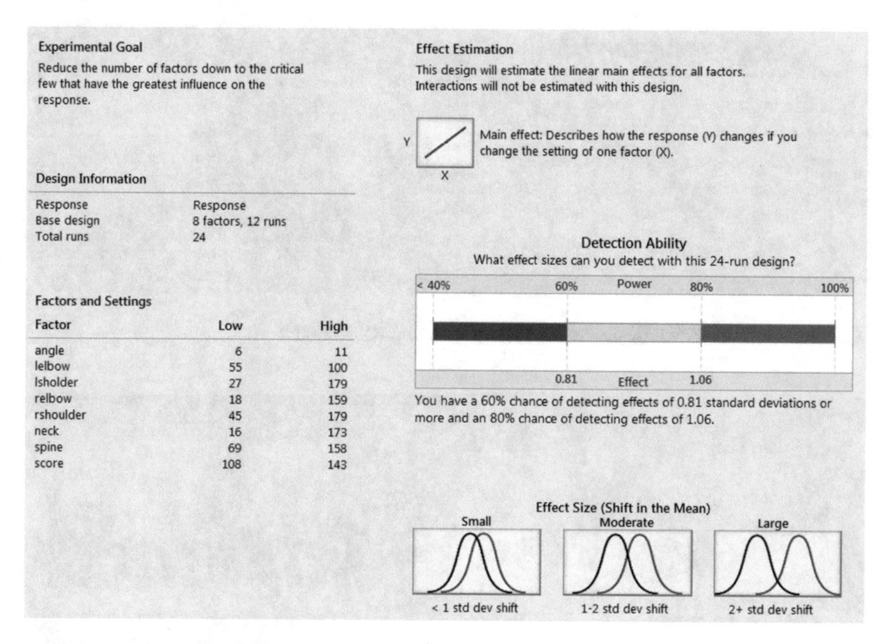

Fig. 4.13 Create screening design

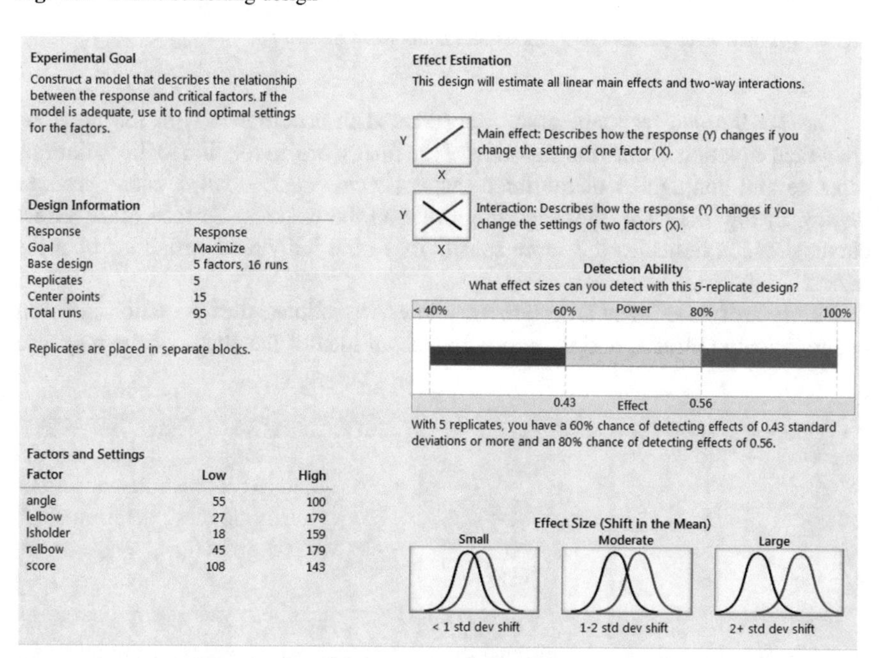

Fig. 4.14 Optimization process in creating modeling design

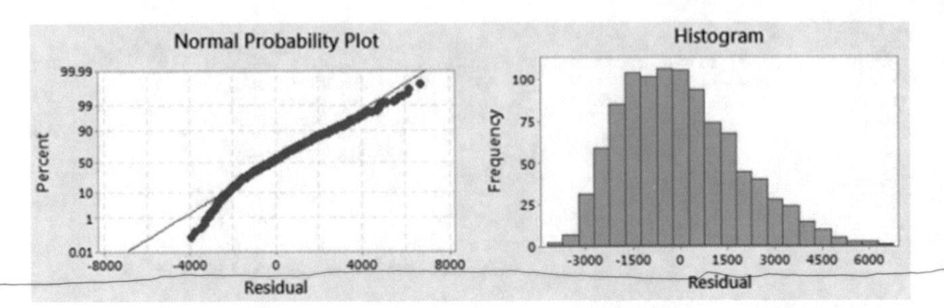

Fig. 4.15 Normal probability plot and histogram using RSM model

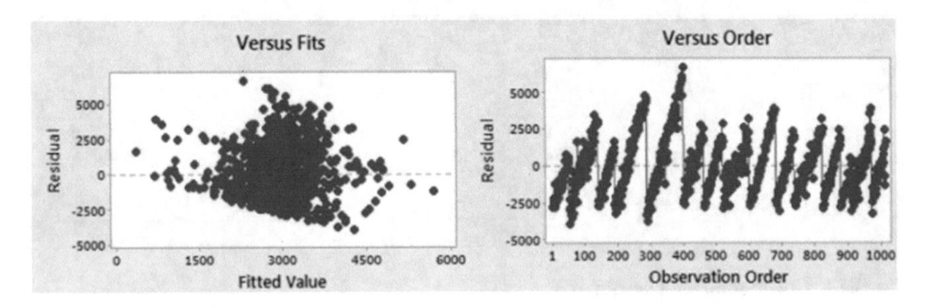

Fig. 4.16 Fitted value versus observation order using RSM model

The 3D Surface response graph, represented in accordance with the previous analytical equation, shows the variation of the final score, in relation to the variations separate and conjugated of the three factors, *spine, neck, elbows, shoulders and angles*. In Fig. 4.17 are illustrated the levels on the response surface (dose–effect curves), which indicates the score sensitivity to the individual variations of these factors.

This type of experiment is designed to allow us to estimate the interaction and even quadratic/cubic effects, and therefore give us an idea of the shape of the response

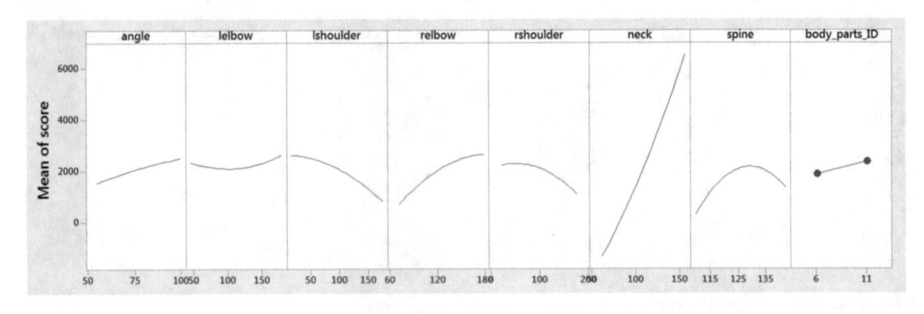

Fig. 4.17 Main effect plot for score

surface we are investigating. The most efficient RSM design for 3 factors and 3-levels is the *Box Behnken design* and the most efficient RSM design for 3 factors and 5-levels is the *Central Composite design*. The value of the estimated correlation coefficient is 0.98 and shows that the current answer (final score calculated by the chosen regression equation) is influenced, separated and/or conjugated, by the variables *spine* and respectively *neck*, in a proportion of 92.5%. The precision of suitability of the chosen model is a percentage that measures the influence of the hazard on the response, in accordance with the chosen response pattern. If this percent has a superior value of significance (5.8%), as is the value of 9.225 in the present case, then the model of the regression equation chosen is suitable.

The goal of Design of Experiments (DOE design) is to get the most information from the fewest amounts of runs. Thus, DOE design is based on specific combinations of:

(1) the # *of Factors* to be tested
(2) the # *of Levels* for each of the factors

The goal of DOE analysis is to achieve reliable, predictable results (Fig. 4.18). For this to happen, four items must be evaluated as part of the analysis:

(1) *P-values:* Significance of Terms in Equation
(2) *R-Square:* Relationship of Inputs to Outputs
(3) $\pm 2 * S$: Predictability of Equation
(4) *Residuals*: Violation of Analysis Assumptions

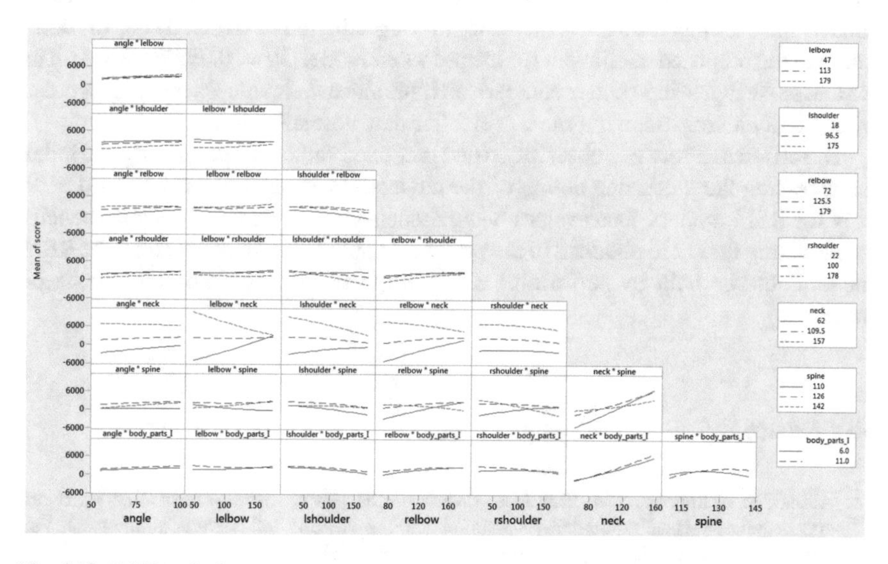

Fig. 4.18 DOE analysis

In order to verify the prediction ability of RSM models, their performance was examined using four well-known statistical methods:

(1) Square root mean error (RMSE),
(2) Mean average percentage error (MAPE),
(3) Sum of square error predictions (SSE),
(4) Determination coefficient (R^2), which are defined as follows [1].

4.4 Conclusion and Future Research Directions

A Response Surface Methodology (RSM) for Medical Rehabilitation based on an Exergaming Platform that uses a Microsoft Kinect Sensors, was presented in this chapter. The development of this application was made in cooperation between a research team from Institute de Telecommunication from Lisbon, Portugal and a research team from the University of Suceava, Romania. In order to find the score for patients with a locomotor leisures or neurodegenerative disorders, we applied the RSM Methodology and we updated the exergame using statistical and mathematical. Our application has demonstrated that the Response Surface Methodology (RSM) is a useful instrument in the prediction of the patients variable scores.

The RSM performance was examined using four statistical methods. For RMS, RMSE, MAPE, and SSE were calculated to be 0.0265, 1.43%, and 0.0084. In addition, for RMSE, MAPE and SSE, the obtained values have confirmed that the RSM models have almost the same superiority in the prediction of the final score as compared to the proposed coefficient for the proposed model ($R^2 = 0.992$ for RSM). The R^2 and MAPE provided better comparison hints since their values were independent of one unit and remained the same even after data normalization.

The difference between observation and predicted values can provide a good index for assessing the prediction ability of the models. Residues range is from −0.38 to 0.34 for RSM models, these values being associated with almost the same capacity in predicting the score obtained in the proposed application. In conclusion, the RSM model could be built by performing a valid regression analysis on the purchased dataset.

References

1. Akbari, S., Mahmood, S.M., Tan, I.M., Hematpour, H.: Comparison of neuro-fuzzy network and response surface methodology pertaining to the viscosity of polymer solutions. J. Pet. Explor. Prod. Technol., **8**(3), 887–900 (2018), ISSN: 2190-0558
2. Active and Assisted Living Programme: ICT for ageing well. Retrieved May 10, 2018, from http://www.aal-europe.eu

3. Chiuchisan, I., Geman, O.: Trends in embedded systems for e-Health and biomedical applications. In: Proceedings of 2016 International Conference and Exposition on Electrical and Power Engineering (EPE), pp. 304–308 (2016). ISBN: 978-1-5090-6129-7
4. Chiuchisan, I., Geman, O., Prelipceanu, M., Costin, H.N.: Health Care System for Monitoring Older Adults in a "Green" Environment using Organic Photovoltaic Devices. Environmental Engineering & Management Journal (EEMJ), 15 (12), pp. 2595–2604 (2016). https://doi.org/10.30638/eemj.2016.286
5. Chiuchisan, I., Geman, O.: An approach of a decision support and home monitoring system for patients with neurological disorders using internet of things concepts. WSEAS Trans. Syst., Issue: Multi-Model. Complex Technol. Syst., **13**, 460–469 (2014a). E-ISSN: 2224-2678
6. Chiuchisan, I., Costin, H.N., Geman, O.: Adopting the Internet of Things Technologies in Health Care Systems. Proceedings of 2014 International Conference and Exposition on Electrical and Power Engineering (EPE), Workshop on Electromagnetic Compatibility and Engineering in Medicine and Biology, pp. 532–535 (2014b). E-ISBN: 978-1-4799-5849-8
7. Chiuchisan, I., Geman, O., Chiuchisan, Iulian, Iuresi, A.C., Graur, A.: NeuroParkinScreen—a Health Care System for Neurological Disorders Screening and Rehabilitation. In: Proceedings of 2014 International Conference and Exposition on Electrical and Power Engineering (EPE), Workshop on Electromagnetic Compatibility and Engineering in Medicine and Biology, pp. 536–540 (2014c). E-ISBN: 978-1-4799-5849-8
8. Da Gama, A., Fallavollita, P., Teichrieb, V., Navab, N.: Motor rehabilitation using Kinect: a systematic review. Games Health J. **4**(2), 123–135 (2015)
9. Dittmar, A., Meffre, R., De Oliveira, F., Gehin, C., Delhomme, G.: Wearable Medical Devices Using Textile and Flexible Technologies for Ambulatory Monitoring, pp. 7161–7164. Engineering in Medicine and Biology Society, IEEE-EMBS (2005)
10. European Commission: Active Ageing Report 2017. Retrieved February 15, 2018, from http://ec.europa.eu/social
11. Geman, O., Postolache, A.P., Chiuchisan, I., Prelipceanu, M.: Jude Hemanth: an intelligent assistive tool using exergaming and response surface methodology for patients with brain disorders, IEEE Access, **7**, 21502–21513 (2019)
12. Geman, O., Toderean, R., Lungu, M.M., Chiuchisan, I., Covasa, M.: Challenges in nutrition education using smart sensors and personalized tools for prevention and control of type 2 diabetes. In: Proceedings of 2017 IEEE 23rd International Symposium for Design and Technology in Electronic Packaging (SIITME), pp. 460–469 (2017). E-ISBN: 978-1-5386-1626-0
13. Geman, O., Hagan, M., Chiuchisan, I.: A novel device for peripheral neuropathy assessment and rehabilitation. In: Proceedings of 2016 International Conference and Exposition on Electrical and Power Engineering (EPE), pp. 309–312 (2016). ISBN: 978-1-5090-6129-7
14. Geman, O., Sanei, S., Costin, H.N., Eftaxias, K., Vysata, O., Prochazka, A., Lhotska, L.: Challenges and trends in Ambient Assisted Living and intelligent tools for disabled and elderly people. In: Proceedings of 2015 International Workshop on Computational Intelligence for Multimedia Understanding (IWCIM), pp. 1–5 (2015). ISBN: 978-1-4673-8457-5
15. Geman, O., Chiuchisan, I., Iuresi, A.C., Chiuchisan, Iulian, Dimian, M. et al.: Intelligent system for a personalized diet of obese patients with cancer. In: Proceedings of 2014 International Conference and Exposition on Electrical and Power Engineering (EPE), Workshop on Electromagnetic Compatibility and Engineering in Medicine and Biology, pp. 528–531 (2014a). E-ISBN: 978-1-4799-5849-8
16. Geman, O., Sanei, S., Chiuchisan, I., Prochazka, A., Vysata, O.: Towards an inclusive parkinson's screening system. In: Proceedings of 18th International Conference on System Theory, Control and Computing, pp. 470–475 (2014b). ISBN: 978-1-4799-4601-3
17. Geman, O., Costin, H.N.: Parkinson's Disease Prediction based on Multistate Markov Models. Int. J. Comput. Commun. & Control. **8**(4), 525–537 (2013a)
18. Geman, O., Costin, H.N.: Tremor and gait screening and rehabilitation system for patients with neurodegenerative disorders. "Buletinul Institutului Politehnic din Iasi" J. Autom. Control. Comput. Sci. Sect., LIX (LXIII) **2**, 43–56 (2013b)

19. Guzsvinecz, T., Szucs, V., Sik Lanyi, C.: Developing movement recognition application with the use of Shimmer sensor and Microsoft Kinect sensor. Stud. Health Technol. Inform. **217**, 767–772 (2015)
20. Kannan, S., Baskar, N.: Modeling and optimization of face milling operation based on response surface methodology and genetic algorithm. Int. J. Eng. Technol. **5**, 959–971 (2013)
21. Kinect: Haas, D., Somphong, P., Jing, Y., et al.: Kinect Based physiotherapy system for home use. Current directions. Biomed. Eng. **1**(1), 180–183 (2015)
22. Kinect: Webster, D., Celik, O.: Systematic review of Kinect applications in elderly care and stroke rehabilitation. J. NeuroEngineering Rehabil. **11**:108–112 (2014)
23. Kinect: Hondori, H., Khademi, M.: A review on technical and clinical impact of microsoft kinect on physical therapy and rehabilitation. J. Med. Eng. **16** (2014)
24. Kinect: Ekstam, L., Johnson, U., Guidetti, S., Eriksson, G.: The combined perceptions of people with stroke and their carers regarding rehabilitation 1 year after stroke: a mixed methods study. BMJ Open **5**(2) (2015)
25. Kinect: Dobkin, B.H.: Rehabilitation after stroke. N. Engl. J. Med., **352**(16):1677–1684 (2005)
26. Kinect: McBean D., Wijck F Van.: Perceptuo-motor control. Applied neurosciences for the Allied Health Professions, pp. 65–109. Churchill Livingstone Elsevier, Britain (2013)
27. Kinect: Archambault, P., Norouzi-Gheidari, N., Kairy, D., Solomon, J.M., Levin, M.F.: Towards establishing clinical guidelines for an arm rehabilitation virtual reality system. Replace, Repair, Restore, Relieve—Bridging Clinical and Engineering Solutions in Neurorehabilitation Biosystems Biorobotics. Switzerland, pp. 263–270. Springer International (2014)
28. Kinect: Tao, G., Archambault, P. S., Levin, M.: Evaluation of a virtual reality rehabilitation system for upper limb hemiparesis. In: International Conference Virtual Rehabilitation (ICVR), pp. 163–165 (2013)
29. Kinect: Lange, B., Chang, C.Y., Suma, E., Newman, B., Rizzo, A.S., Bolas, M.: Development and evaluation of low cost game-based balance rehabilitation tool using Microsoft Kinect sensor. Engineering in Medicine and Biology Society, EMBC, pp. 1831–1834 (2011)
30. Mintal, F.A., Szucs, V., Sik-Lanyi, C.: Developing movement therapy application with Microsoft Kinect control for supporting stroke rehabilitation. Stud. Health Technol. Inform. **217**, 773–781 (2015)
31. Myers, R.H., Montgomery, D.C., et al: Response surface methodology: a retrospective and literature survey. J. Qual. Technol. **36**(1):53–77 (2018)
32. Parry, I., Carbullido, C., Kawada, J., Bagley, A., et al.: Keeping up with video game technology: objective analysis of Xbox Kinect™ and PlayStation 3 Move™ for use in burn rehabilitation. Burns, **40**(5), 852–859 (2014)
33. Palumbo, F.: Ambient intelligence in assisted living environments. Ph.D. Thesis, Universita degli Studi di Pisa, Dipartimento di Informatica, Dottorato di Ricerca in Informatica (2016)
34. Postolache, O.: Project: smart sensors and tailored environments for physiotherapy. Retrieved May 5, 2018 from: https://www.it.pt/Projects/Index/3223
35. Postolache, O., Viegas, V.V., Dias Pereira, J.M., Girao, P.M.: Smart Sensors Architectures for Vital Signs and Motor Activity Monitoring. Chapter in Advanced Interfacing Techniques for Sensors Measurement Circuits and Systems for Intelligent Sensors. Springer International Publishing, Cham, Switzerland (2017)
36. Postolache, O., Postolache, G., Carvalho, H.C., Catarino, A.C.: Smart Clothes for Rehabilitation Context: Technical and Technological Issues. Chapter in Sensors for Everyday Life Healthcare Settings. Springer Berlin Heidelberg, Berlin (2016)
37. Postolache, O., Dias Pereira, J.M., Ribeiro, M.R., Girao, P.M.: Assistive Smart Sensing Devices for Gait Rehabilitation Monitoring. Chapter in ICTs for Improving Patients Rehabilitation Research Techniques. Springer International Publishing, Berlin Heidelberg (2015)
38. Postolache, O., Girao, P.M., Postolache, G.: Pervasive Sensing and M-Health: Vital Signs and Daily Activity Monitoring. Chapter in Pervasive and Mobile Sensing and Computing for Healthcare. Springer International Publishing, Heidelberg (2012a)
39. Postolache, G., Girao, P.M., Postolache, O.: Requirements and Barriers to Pervasive Health Adoption. Chapter in Pervasive and Mobile Sensing and Computing for Healthcare—Technological and Social Issues. Springer International Publishing, Heildelberg (2012b)

40. Postolache, O, Girao, P.M., Dias Pereira, J. M.: Water Quality Assessment Through Smart Sensing and Computational Intelligence. Chapter in New Developments and Applications in Sensing. Technology Springer International Publishing. Berlin Heidelberg (2011a)
41. Postolache, O., Dias Pereira, J. M., Girao, P.M.: Underwater Acoustic Source Localization and Sounds Classification in Distributed Measurement Networks. Chapter in Advances in Sound Localization, Pawel Strumillo, In-Tech, Wien (2011b)
42. Postolache, O., Dias Pereira, J.M., Girao, P.M., Postolache, G.: Distributed Smart Sensing Systems for Indoor Monitoring of Respiratory Distress Triggering factors. Chapter in chemistry, emission, control, radioactive pollution and indoor air quality. Intech, In-Tech, Rijeka (2011c)
43. Postolache, O., Girao, P.M., Pinheiro, E.C., Postolache, G.: Unobtrusive and Non-invasive Sensing Solutions for on-line Physiological Parameters Monitoring. Chapter in Wearable and Autonomous Biomedical Devices and Systems for Smart Environment. Springer International Publishing, Berlin (2010)
44. Ribeiro, J.M., Postolache, O., Girao, P.M.: A Novel Smart Sensing Platform for Vital Signs and Motor Activity Monitoring. Chapter in Sensing Technology: Current Status and Future Trends. Springer International Publishing, Heidelberg (2014)
45. Swanson, L.R., Whittinghill, D.M.: Intrinsic or extrinsic? using videogames to motivate stroke survivors: a systematic review. Games Health J. **4**(3), 253–258 (2015)
46. Webster, D., Celik, O.: Systematic review of Kinect applications in elderly care and stroke rehabilitation. J. Neuroeng. Rehabil. **3** (2014)
47. Wittland, J., Brauner, P., Ziefle, M.: Serious games for cognitive training in ambient assisted living environments—a technology acceptance perspective. In: Proceedings of 15th Interact 2015 Conference, LNCS Vol. 9296, Springer International Publishing, pp. 453–471(2015)
48. World Health Organization: WHO guidelines on integrated care for older people (ICOPE). ISBN: 9789241550109 (2017)
49. Yonghee, Y., Sangmun, S.: Job stress evaluation using response surface data mining. Int. J. Industr. Ergonom. **40**, 379–385 (2010)

Additional Reading Section (Resource List)

50. Allen, D.M.: Mean square error of prediction as a criterion for selecting variables. Technometrics, 13 (1971), ISSN:469-475
51. Allen, D.M.: The relationship between variable selection and data augmentation and a method for prediction. Technometrics, **16**(1974), ISSN:125-127
52. Ahanathapillai, V., Amorx, J., James, C.: Assistive tech-nology to monitor activity, health and wellbeing in old age: the wrist wearable unit in the USEFIL project. Techno. lDisabil. **27**, 17–29 (2015)
53. Bo, A. P. L., Hayashibe, P. Poignet: Joint angle estimation in rehabilitation with inertial sensors and its integration with Kinect. In: Lovell, N. (ed) Engineering in Medicine and Biology Society Annual International Conference, pp. 3479 – 83 (2011)
54. Box, G.E.P., Draper, N.R.: Empirical Model-Building and Response Surfaces. Wiley, New York (1987)
55. Bolandzadeh, N., Kording, K., Salowitz, N., Davis, J.C., Hsu, L., Chan, A., et al.: Predicting cognitive function from clinical measures of physical function and health status in older adults. PLoS ONE **10**, e0119075 (2015)
56. Bridenbaugh, S.A., Kressig, R.W.: Motor cognitive dual tasking: early detection of gait impairment, fall risk and cognitive decline. Z. Gerontol. Geriatr. **48**, 15–21 (2015)
57. Cho, K.H., Lee, W.H.: Virtual walking training program using a real-world video recording for patients with chronic stroke: a pilot study. Am. J. Phys. Med. Rehabil. **92**, 371–380 (2013)
58. Chang, I.S.J., Boger, J. Qiu, J. Mihailidis, A.: Pervasive Computing and Ambient Physiological Monitoring Devices. Assistive Technologies in Smart Environments for People with Disabilities. Boca Raton, FL: CRC Press (2015)

59. de Joode, E., van Heugten, C., Verhey, F., van Boxtel, M.: Efficacy and usability of assistive technology for patientswith cognitive deficits: a systematic review. Clin. Rehabil. **24**, 701–714 (2010)
60. Ijsselsteijn, W.A., Nap, H.H., De Kort, Y., Poels, K.: Digital game design for elderly users. In: Proceedings of the 2007 Conference on Future Play, ACM press, New York, NY, pp. 17–22 (2007)
61. Ienca, M., Jotterand, F., Vica, C., Elger, B.: Social and assistive robotics in dementia care: ethical recommendations for research and practice. Int. J. Soc. Robot. **8**, 565–573 (2016)
62. Joshi., S., Sherali, H.D., Tew, J.D.: An Enhanced Response Surface Methodology (RSM) Algorithm Using Gradient Deflection and Second-Order Search Strategies. Comput. Oper. 2 (7/8), pp. 531–541 (1998)
63. Khuri, A.I., Cornell, J.A.: Response Surfaces, 2nd edn. Marcel Dekker, New York (1996)
64. Khosravi, P., Ghapanchi, A.H.: Investigating the effec-tiveness of technologies applied to assist seniors: asystematic literature review. Int. J. Med. Inform. **85**, 17–26 (2016)
65. Laver, K.E., George, S., Thomas, S., Deutsch, J.E., Crotty, M.: Virtual reality for stroke rehabilitation. Cochrane Database Syst. Rev. **9** (2011)
66. Myers, R.H., Montgomery, D.C.: Response Surface Methodology: Process and Product Optimization Using Designed Experiment. A Wiley-Interscience Publication (2002)
67. Montgomery, D.C., Peck, E.A., Vining, G.G.: Introduction to linear regression analysis, 3rd ed. Willey, New York (2001)
68. Myers, R.H.: Classical and Modern Regression with Applications, 2nd edn. Duxbury Press, Boston (1990)
69. Neddermeijer, H.G., van Oortmarssen G.J., Piersma N., Dekker R.: A Framework for Response Surface Methodology for Simulation Optimization Models. Proceedings of the 2000 Winter Simulation Conference, pp. 129–136 (2000)
70. Pasch, M., Bianchi-Berthouze, N., Van Dijk, B., Nijholt, A.: Movement-based sports video games: investigating motivation and gaming experience. Entertain. Comput. **1**, 49–61 (2009)
71. Russel, S.J., Norvig P.: Artificial Intelligence: A Modern Approach. Prentice Hall (1995)
72. Simpson, J.: Challenges and trends driving telerehabilitation. Telerehabilitation, pp. 13–27. Springer-Verlag, London (2013)
73. Taguchi, G.: System of Experimental Design: Engineering Methods to Optimize Quality and Minimize Cost. UNIPUB/Kraus International, White Plains, NY (1987)
74. Zhang, Q., Su, Y., Yu, P.: Assisting an elderly with earlydementia using wireless sensors data in smarter safer home. In: 15th IFIP WG 8.1 international Conference on Informatics and Semiotics in Organisations, ICISO 2014, Springer New York LLC, pp. 398–404 (2014)

Chapter 5
An Integrated System for Improved Assisted Living of Elderly People

Imad Alex Awada, Irina Mocanu, Alexandru Sorici and Adina Magda Florea

Abstract The number of elderly people (aged 60 years or over) is increasing significantly. Moreover, this happens in the context of increasing well-being costs and decreasing caregiver availability. Therefore, technology must create assisted living solutions that support elderly in their daily activities and ensure their continuous health-monitoring, safety and social integration while maintaining an acceptable degree of independence. In this context we present the CAMI system—an intelligent system that provides health and home monitoring, supervised physical exercises and interaction between the user and the system is performed through a multimodal interface. The supervised physical exercises are recommended based on the current medical parameters of the user. The results performed by the user are presented in an interactive way using a feedback module. The multimodal interface accepts voice and gesture-based commands and is adapted to the device and to the user profile and preferences.

Keywords Ambient assisted living · Elderly people support · Health monitoring · Multimodal interface · Supervised physical exercises · User profile

I. A. Awada (✉) · I. Mocanu · A. Sorici · A. M. Florea
Computer Science Department, University Politehnica of Bucharest, Bucharest, Romania
e-mail: awadaalex@hotmail.com

I. Mocanu
e-mail: irina.mocanu@cs.pub.ro

A. Sorici
e-mail: alexandru.sorici@cs.pub.ro

A. M. Florea
e-mail: adina.florea@cs.pub.ro

H. Costin et al. (eds.), *Recent Advances in Intelligent Assistive Technologies: Paradigms and Applications*, Intelligent Systems Reference Library 170,
https://doi.org/10.1007/978-3-030-30817-9_5

5.1 Introduction

The increasing ageing population demands for solutions that help in the independent, healthy and risk-free life of the elderly and prevent their social isolation. Ambient Assisted Living (AAL) use Information and Communication Technologies (ICT) in a living or working environment to enable elderly people or people with special needs to remain active, socially connected and live independently by using pervasive computing, ubiquitous communication, and intelligent user interfaces.

Initial project calls within the AAL initiative have focused on specific aspects and needs of the senior population (e.g. management of chronic conditions, advancement of social interaction for elderly people, supporting occupation of in life of older adults, living actively and independently at home). This has led to the development of many ICT solutions targeting these specific needs of the elderly.

However, the most recent call sees a shift of attention towards developing integrational platforms/AAL packages which offer a comprehensive set of functionalities (such as the ones listed above) into a single system. The focus falls on integrating existing frameworks into a unitary solution, for which a viable business plan can be developed.

Aiming to live up to this perspective, we have developed CAMI, an Artificial Intelligence based System for Self-Management and Sustainable Quality of Life in AAL. CAMI proposes a fully integrated AAL solution at the overlap of telecare and health, smart homes and robotics, by offering services for social, health and home care, as well as compensation for reduced mobility (see Fig. 5.1 for service list).

The target group of CAMI is older adults in general and older adults with a risk for cardiovascular diseases, diabetes or mild cognitive impairment in particular. CAMI aims to provide flexible, scalable, and individualized services that enables self-monitoring of this group.

To support this level of integration, the CAMI system is designed as a highly modular and loosely-coupled cloud-based architecture that integrates powerful ambient assisted living features (health and home monitoring, supervised physical exercises, fall detection) and a multimodal interface that ensures a smooth access to the different functionalities of the system.

This paper describes three main components of the CAMI system: the context-aware decision-making module, the resulting multimodal interaction with the user and the means to stimulate the user's physical activity. These showcase the range of integration capabilities of CAMI. The rest of the paper is organized as follows. In Sect. 5.2 we position CAMI with respect to previous work in healthcare and health monitoring systems which attempted to package more functionalities into the one system. We analyze the range of functionalities and the architectural solutions and show how CAMI has a more flexible approach and a more comprehensive list of features. Section 5.3 gives details about the range of functionalities available in the CAMI System. Section 5.4 describes the architecture which allows for this degree of integration, by building on a principle of loosely coupled micro-services, communicating via a message processing system and RESTful Web APIs. In Sect. 5.5

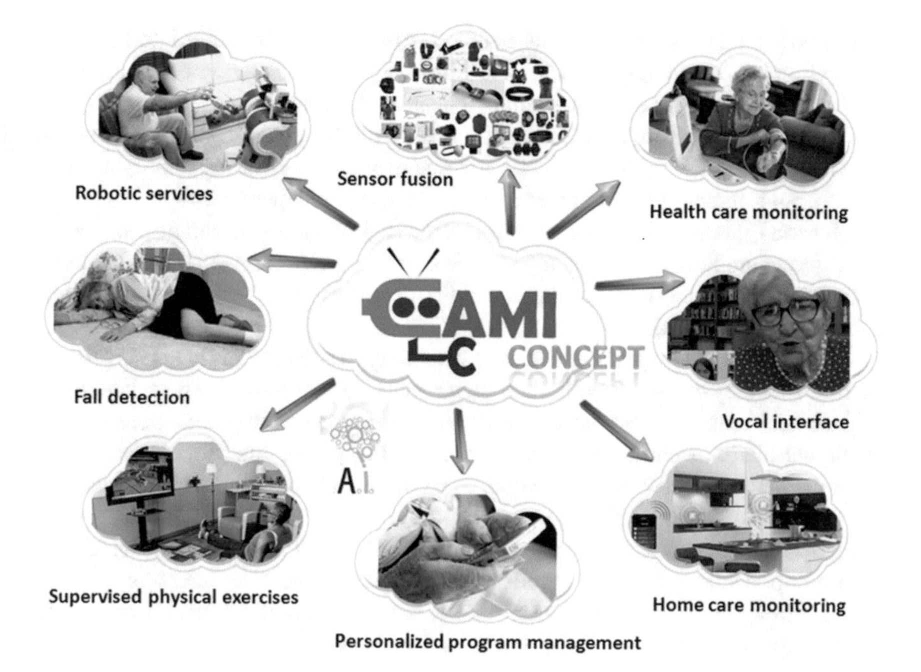

Fig. 5.1 CAMI concept and services

we then go on to detail the functionality of the framework that underlies the intelligent health analysis and notification management service in the CAMI System. We explain how it is instantiated in the case of CAMI. Section 5.6 describes the multimodal interface available in the CAMI System, giving details about the different modules of the adaptive interface, voice interaction capabilities and interaction via a humanoid robot. Section 5.7 gives details about the way in which the CAMI System accustoms for the need of compensating for reduced mobility in elderly citizens. It presents the functionality of the physical exercise monitoring module and the way it is integrated into the CAMI System. Finally, Sect. 8 provides conclusions in terms of the usability of the system based on preliminary field trials and considerations for future work.

5.2 Related Work

The necessity to provide an encompassing approach to care for the elderly at home and to creating the ability for their sustained autonomy of life has opened new dimensions in the eHealth environment through ICT.

As mentioned previously, the AAL community has recently put a lot of accent towards encouraging systems that focus on packaging together several solutions that have been developed throughout the years related to health management and aging at home.

Moreover, there is an increasing demand from patients for using internet-based applications in the management of their diseases as well as having better access to their care providers [1] and maintaining a healthy lifestyle.

From this perspective, different solutions were produced aiming at allowing users to live longer in their favorite medium (own house), while ensuring a safe and comfortable environment [2] with an acceptable degree of independence.

Persona [3] is an older, open-standard technological platform, which allows building different ambient assisted living services, such as supporting the social inclusion of the elderly people, monitoring their daily activities, helping them to feel more safe, secure and confident.

The **inCASA** [4] project aims to use technology to protect elderly and to prolong the time that they can live well in their own homes. The system integrates health and environment sensors to collect and analyze data in order to implement customized intelligent alerts or communication services and to profile the users behaviour. A smart personal platform makes the data of the user available for the care services.

The **healthy@work** [5] project targets to improve the quality of life and well-being of older employees. The system is composed of mobile and server-based software components; and some mobile sensor-systems. The user interacts with the system through a mobile application that creates a personalized health promotion program that contains small daily activities, inputs and monitoring (e.g. change the daily behavior to a healthier one).

In order to ensure a healthy life at the workplace for the older adults, the Wellbeing [6] project offers a web-based platform that combines four modules: workplace ergonomics, stress management, physical exercises and nutritional balance modules. The different modules of the platform use a 3D sensor and an RGB camera to track the users. In a case of an unhealthy situation (e.g. wrong sitting position, high stress level) the platform notifies the user and provides information to the user on how to change the situation to a healthy one.

PersonAAL project [7] offers a technology that allows the elderly people to live more time in their home through an intuitive web application. The elderly person can receive personalized and context-dependent assistance through the application while being inside their home which improves the quality of life on the one hand and decreases the healthcare delivery cost on the other (e.g. motivate the elderly person in his/her daily routines to achieve a specific goal such as having a healthier lifestyle by eating healthy food together with the office coworkers).

The **EldersUP!** [8] project targets to create connections between the elderly people and start-ups or small companies which helps the seniors to have an active social life and make them feel useful for the society for themselves and the society from a part. Furthermore, it helps the small companies and the start-ups to transfer the valuable experiences of the elderly to their employees. The system monitors the cognitive abilities and engagement level of the end user and adapts the interface and content of the workspace to the cognitive conditions of the user. Even though the EldersUp! project does not offer any health or environment monitoring services, we analyzed it, because of its adaptive interface and its role in improving the quality of life of the elderly people by protecting them from failing into apathy after their retirement.

In analyzing the previous solutions, we notice that neither one integrates a module for supervised physical exercises at home, which is an important module to help the elderly people have an active and healthy life. Solutions such as inCASA and healthy@work do not offer options for multimodal user interaction (e.g. using voice commands). While they relate to several AAL aspects, the focus of the EldersUP! and healthy@work projects does not lie on monitoring solutions for personal homes. Lastly, the PersonAAL and EldersUP! projects are the only ones among the set of reviewed systems that take into consideration adaptive capabilities for their user interfaces.

All the mentioned solutions (commercial or research projects) have limited functionalities and do not cover all the functionalities that we want to implement in the CAMI system. Furthermore, from an architectural perspective, the reviewed solutions are bound to one dominant technology that underlies the implementation. This means that future integrations of new modules to these systems are predicated on an effort to develop these modules in the technology of the systems. In the system we propose, modules are conceived as micro-services that are loosely coupled and that communicate via message brokers (e.g. RabbitMQ) and RESTful APIs. This not only increases the flexibility of the system and the range of modules it can integrate, but it also constitutes a logistical advantage, since it allows members in a project consortium to work more independently on their own module, as long as the interactions with the other modules are well described using API definition files.

5.3 CAMI Functionalities

The services offered by CAMI ecosystem address both healthy individuals as well as those with age-related impairments. CAMI solution reconciles the increased demand for care in the current aging society with limited resources by supporting an efficient and sustainable care system. One of the main advantages of the CAMI solution is that it is configurable and adaptable according to the services requested by the user. Moreover, it can be extended at a later time, with more functionalities and connected devices, as the need arises. The overall CAMI concept and offered services to the user are presented in Fig. 5.1.

The concept and functionalities are achieved through a modular and flexible architecture, integrating several technologies, platforms, and artificial intelligence techniques. These functionalities include health parameter monitoring, home environment monitoring, stimulating user's physical activity, fall detection, issuing advices and reminders, daily activity planning, communicating with caregivers and interaction with the system through a robotic platform.

CAMI offers a set of different services to the user. The main functionalities of the CAMI system are presented below:

- **Health monitoring**: the system performs regular monitoring of health parameters and also continuous monitoring of health parameters during physical exercises.

The health parameters that are monitored are: blood pressure, weight, heart rate, number of steps and sleep activity. Data recorded is stored and correlated with the user profile and is available to caregivers. Also, it is used for selecting the level and intensity of physical exercises. Some health parameters are continuous monitoring—for example heart rate—during the physical exercises such that the user remains in her/his comfort and low risk zone.

- **Home and environment management**: the system controls the environment parameters in the user premises. Some of the parameters are used in connection with the home appliances to increase the comfort of the user. Others are used to control the environment, e.g. remote switching of lights. The CAMI system permanently measures and monitors: environment temperature, motion, illumination, if windows are opened or closed and also through plug switch devices can be turned on or off.
- **Fall alarm**: detected falls are sent to formal or informal caregivers. For fall detection we used only wearable fall detection sensor (e.g. Vibby Oak and its IoT gateway, Vibby Leaf). Vibby Oak fall detection sensors are designed to detect automatically heavy-dangerous falls of its wearer lying on the floor with or without manual activity. The device also has a manual trigger to push the alarm off, if the user has recovered successfully. The Vibby Oak sensors communicate via radio signals with a gateway called the Vibby Leaf base.
- **User interaction through a robotic platform**: A robotic platform in CAMI is dedicated to provide assistance for people who have difficulties in moving or are incapable of doing so, or even to provide psychological wellbeing. For example: the robot displays relevant information to the user, e.g., the latest health status updates; it searches and detects the user, recognizes him/her, and moves towards the person when there are notifications from the CAMI System that require the end user's attention, e.g., an unacknowledged reminder for taking prescribed medication. Thus, the CAMI system provides the following three components:

  - people detection—the robot needs to know where the person is in the room;
  - people recognition—the robot has to be able to recognize the person so as to provide him/her with the information specific to the user, by accessing this information in the CAMI cloud;
  - navigation in the environment—once the person is detected, the robot has to navigate to approach the person, display relevant information, allow interaction of that person with the robot, both touch and voice interaction; in order to go near the detected person, the robot has to do a mapping of the environment, localizing itself in the environment and then navigate between its present location and the location of the detected person.

  These components are based on advanced artificial intelligence algorithms and techniques. They have the advantages that can be installed and configured on several robotic platforms, including semi-humanoid ones or telepresence.

- **Report to health professionals**: Health parameters acquired can be provided to formal and informal caregivers. The vital parameters associated with a person are visualized through a multimodal interface that can be accessed through voice commands and touch based gestures. The health parameters or alerts are provided in specific ways for different type of caregivers (formal or informal).
- **Stimulating User's Physical Activity**: In case of low physical activity (detected based on the measured health parameters: e.g. number of steps) the user is advised to increase her/his level of physical exercise. The exercises are performed interactively, as a game, specifically designed for elderly persons using the Kinect v2 sensor. The type of the exercise, the intensity level of the recommended physical exercises and the duration of the exercise are selected based on the vital parameters of the user measured by the health monitoring component. The physical exercises are presented to the user using an avatar. The training avatar performs different physical exercises and the user must reproduce his movements. Then the user's movements are compared with the movements of the avatar and the results are saved in order to be analyzed by the caregiver. At the end of the exercise, the user receives a score that reflects the correctness of the performed exercises. Also, a feedback about the performed exercises is shown to the user. The feedback is given through multimodal interface. It includes an avatar that replay the exercise performed by the user. Also, the review includes the mistakes committed by the user during the session. A score and some suggestions that improve user's performance are also included into the user's feedback. After performing an exercise, another exercise will be recommended to the user—the type of the next exercise will be selected based on the current pulse of the user and also based on its face emotions.
- **Personalized, intelligent and dynamic program management**: allows the introduction of personal data about the user: medication plan, daily, weekly and month program planning, exercise planner, record of data obtained from sensors, including medical data, interactions with formal and informal caregivers. Depending on various conditions and actual recorded data, it is compared what has been achieved to what has been planned and is able to dynamically adjust the user program. The system has the ability to view/create/modify activities grouped into four categories: (i) medication reminders; (ii) physical exercises; (iii) health measurement requests and (iv) personal (e.g. other end-user activities such as visits to friends, various appointments, etc.). For each type of activity, the user has the option of defining either a one-time event, or a recurrent one. When defining a recurrent activity, the user can specify: (i) an event that recurs several times a day (e.g. daily at 08:00 and 18:00) and (ii) an event that recurs several times a week, specifying the given weekdays (e.g. every Monday, Wednesday and Friday at 12:00). Also, the CAMI system offers synchronization of the local calendar with a Google Calendar instance that belongs to the end user.
- **Voice interaction module to allow multimodal interaction**: It is recognized that some traditional interfaces can be overly difficult to use by elderly citizens. We therefore consider providing means for multimodal interaction with the CAMI System via voice commands. We evaluate the appeal of this option in a user study.

5.4 CAMI Architecture

CAMI is designed as a highly modular, micro-service-based information system, where individual modules and services can enact their own lifecycle, while at the same time responding to events and service requests from other modules.

Figure 5.2 presents a block diagram depicting the separation of CAMI into its main logical components: the Sensor Unit, the CAMI Gateway, the CAMI Cloud and the CAMI Multimodal Interface.

The Sensor Unit includes all the physical sensors that monitor the health status of a person, the situation in his/her environment (home monitoring sensors), as well as the sensor setup required for performing indoor physical exercises.

The CAMI Gateway is implemented on top of the Eclexys SNG-Gateway and is a service that constructs a representation for all health measurements and environment events generated by the Sensor Unit and forwards them for interpretation and handling to the CAMI Cloud. Furthermore, it is the place where the module for running and evaluating the execution correctness of physical exercises is deployed.

The CAMI Cloud is composed of a set of micro-services which define the functionality of the CAMI system. Apart from the infrastructure (e.g. the Event Stream Manager) that enables the micro-services to interact with each other, the CAMI clouds hosts services that enable user account configuration and an analysis of the measurements and events received from the CAMI Gateway.

The Decision Support Service (DSS) is responsible for issuing notifications or reminders to the end-user based on the observed information.

The CAMI multimodal interface is responsible for displaying information about the current health status as well as the notifications and reminders issued by the DSS

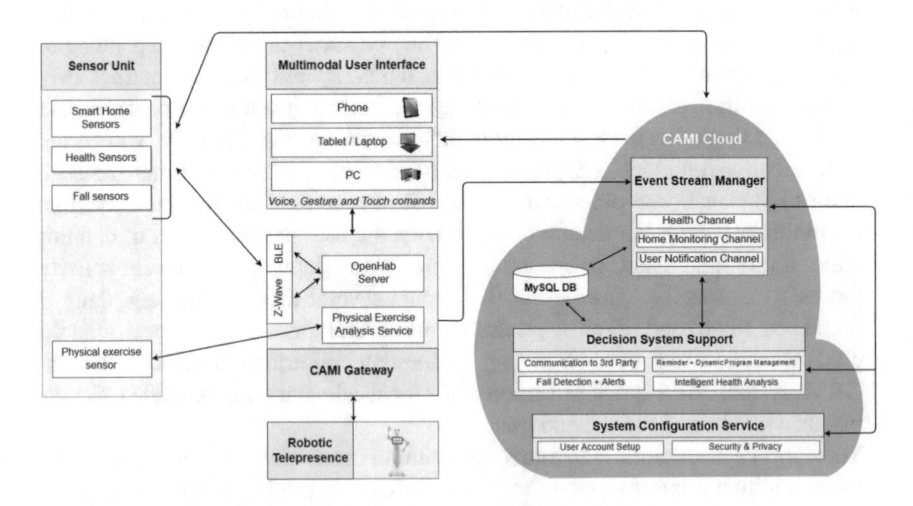

Fig. 5.2 Block diagram of the CAMI system

to the user. This happens across different devices (e.g. mobile, tablet, laptop, PC, etc.) and multiple interaction methods (e.g. touch based, voice based, etc.).

5.4.1 CAMI Sensor Unit and CAMI Gateway

The typical target user of the CAMI system are elderly people aged 55–75, that want or need to monitor their health condition and everyday activity.

Consequently, the sensors already integrated in the system include both medically certified, as well as general consumer-grade devices that can track health parameters and monitor activity.

From a technical perspective, the CAMI system integrates with two types of sensors:

- Sensors that emit data only locally (e.g. based on Personal Area Network wireless protocols such as Bluetooth Low Energy and Z-Wave) and which are routed through the CAMI Gateway to the CAMI Cloud.
- Sensors that push their data directly to a cloud service (e.g. Fitbit, Apple Healthkit, Google Fit) from where it is retrieved directly from the CAMI Cloud.

The CAMI Gateway enables the system to retrieve data from any Bluetooth Health Device (HDP-BT) compatible device, as well as from A&D Medical [9] BLE (Bluetooth 4.0 Low Energy) devices.

Additionally, the gateway can pair to and communicate with Z-Wave compatible home monitoring devices, enabling both data collection and control via the openHAB platform [10].

A default sensor line-up for the CAMI System includes devices such as:

- Health Measurement Devices
  - A&D UA-651 BLE blood pressure meter,
  - A&D UA-352BLE weight scale,
  - Withings WS30 weight scale—device that sends data directly to the cloud,
  - Fitbit bracelet—used to collect information about heart rate, number of steps and sleep activity.
- Home Monitoring Devices
  - Fibaro Temperature and Motion Sensor FGMS-001,
  - Philio PST02 Slim Multisensor (presence detection, door/window open/closed detection, temperature and illumination),
  - Fibaro FGWP101 Metered Wall Plug Switch—used to actuate (turn on/off) devices plugged into the switch (e.g. lamps).

Apart from openHAB, the CAMI Gateway implements a forwarding service which listens for received health measurements and events from the home monitoring sensors. The forwarding service converts all information according to a well-defined format and sends it via an HTTP call to the CAMI Cloud.

We describe the message format and the way in which they are received cloud-side in the following section.

5.4.2 CAMI Cloud

The CAMI Cloud hosts a set of micro-services that provide information collection, storing, analysis and sharing. Each micro-service is deployed in its own Docker [11] container, making it easy to manage the lifecycle of the service.

From a data collection and integration point of view the CAMI Cloud fulfills two essential tasks:

- Collects health measurement and home monitoring event information from the CAMI Gateway (as described previously).
- Actively subscribe to 3rd party APIs to collect information from devices that publish their data directly to 3rd party cloud (e.g. Withings weight scale, FitBit). The System Configuration Service (see Fig. 5.2) offers the necessary information to bind the CAMI Cloud user account to 3rd party applications that he/she may use.

Figure 5.2 depicts an Event Stream Management service that handles all information coming into the CAMI cloud. The Event Stream Manager is implemented using a set of RabbitMQ [12] channels which route messages to the micro-services that want/need to listen to them.

There are separate channels for health measurements and home monitoring events. Other micro-services, such as the Decision Support System (DSS), can subscribe to these channels to receive the messages exchanged over them.

The CAMI Cloud defines an *insertion API* that specifies a URI and a message form where HTTP requests can be made to insert new health or home monitoring data into the Event Stream Manager. The API is described using a Swagger API definition file [13], which makes it easy for developers to build clients that exploit the API in an easy manner. The insertion API defines payload formats for both measurements and home monitoring events (motion sensor triggers, open/closed state of a door/window sensor). For the home monitoring events, special attention is also given to meta-information (e.g. the source of the information, the certainty of the sensor value, timestamp and temporal validity of the event).

The DSS subscribes to updates of all health and home monitoring data and performs analyses such as: abnormal health parameter value detection depending on the context (e.g. increased *resting* heart rate, as opposed to increase due to performing a physical exercise), issuing of reminders for medication or mandatory measurements depending on home monitoring events (e.g. send a reminder for a morning weight measurement, if this is the first activation of the bathroom motion sensor this morning).

The notifications and reminders of the DSS are inserted back into the Event Manager Service in a *notification* queue. A push-notification micro-service is subscribed to messages arriving on this queue and sends the notifications and reminders for consumption by the mobile client interface.

We describe the workings of the DSS via an implementation based on the CONSERT Middleware [14], which is presented in the following section.

The way in which end-users (both elderly people and their caregivers) interact with the reasoning output of the CAMI System to observe their health status, measurement history and received notifications/reminders is presented in Sect. 5.6.

5.5 CAMI Intelligent Decision Making

The CAMI Decision Support System offers functionality dealing with communication to 3rd party health monitoring systems, health analysis, reminder and notification management, dynamic management of weekly activity schedule, as well as alert sending in case of detected falls. We bring focus to two of the mentioned aspects, namely, *health analysis* and *reminder and notification management.*

Health Analysis
CAMI continuously monitors health status parameters such as weight, blood pressure, heart rate, step count and sleep activity. Depending on the needs of the elderly person, sensor that measure blood oxygen levels or blood glucose levels can be easily integrated into the system.

One of the important tasks of CAMI is to issue timely notifications to both the elderly person and his/her caregivers when the monitored health parameters deviate substantially from the typical norm.

The CAMI DSS will send alert notifications in the following cases:

- The systolic and diastolic values are noticeably lower or higher than the standards defined by the American Heart Association (AHA) [15].
- The daytime pulse values are noticeably higher than the standards defined by the AHA **and** the user is not in the middle of an exercise.
- The daytime pulse values are noticeably lower than the standards defined by the AHA **and** the user is not sleeping.
- The user has made less than 6000 steps from the start of the day until 7 PM, local time.
- The user has two successive morning weight measurements with an absolute difference of more than 2 kg.

Reminder and Notification Management
We have already mentioned some cases in which the user receives notifications based on deviations from typical health measurements. In addition to these events, the CAMI System sends reminders for scheduled medication intake or for mandated

health measurements. However, mandated health measurements are not defined based on a fixed moment in time, but rather relative to other user activities (e.g. waking up, eating).

For example, in one considered scenario, the system determines that the user has woken up based on the triggering of a motion sensor installed in the bathroom. The system waits for 10 min for mandatory early morning weight and blood pressure measurements to arrive. This is to say, that the CAMI System considers that the user remembers by himself to take the required measurements, without having to be reminded right after triggering the sensor.

If no measurement is received within 10 min, the CAMI System sends a reminder for the required measurements and awaits the reception of the values for another 10 min.

If no measurements are received this second time around, a notification is sent to the caregivers.

Any measurements received after this time will be recorded, but not validated.

From this simple scenario, it becomes apparent that the notification management system has to be able to reason over the order and distance in time of events, as well as the lack of events over a given period of time. The system must additionally reason over data collected from different sources, e.g. health measurement devices and environment monitoring sensors, to put situations into context. This type of reasoning and functionality is enabled by the CONSERT Middleware.

5.5.1 CONSERT Middleware

CONSERT is a Context Management Middleware [14] which offers support for expressive context modeling and reasoning on one hand, flexible deployment options and adaptable context provisioning mechanisms, on the other hand.

The middleware is structured as a hierarchically ordered collection of logical work groups (called Context Management Units) which consist of service units (agents) handling acquisition, inference, query management and usage of context information. Each inference agent is capable of performing reasoning over context information (e.g. derivation of higher level information, consistency management) using a semantic complex event processing engine called the CONSERT Engine [16, 17].

Figure 5.3 displays a block diagram of the CONSERT Middleware architecture. Notice that there are 5 conceptual agent (service) types defined.

A set of five agent types is defined. They constitute what are called Context Management Units (CMUs). For a given application, multiple CMUs can be deployed and organized in different ways for managing and provisioning context to the applications. These agent types are:

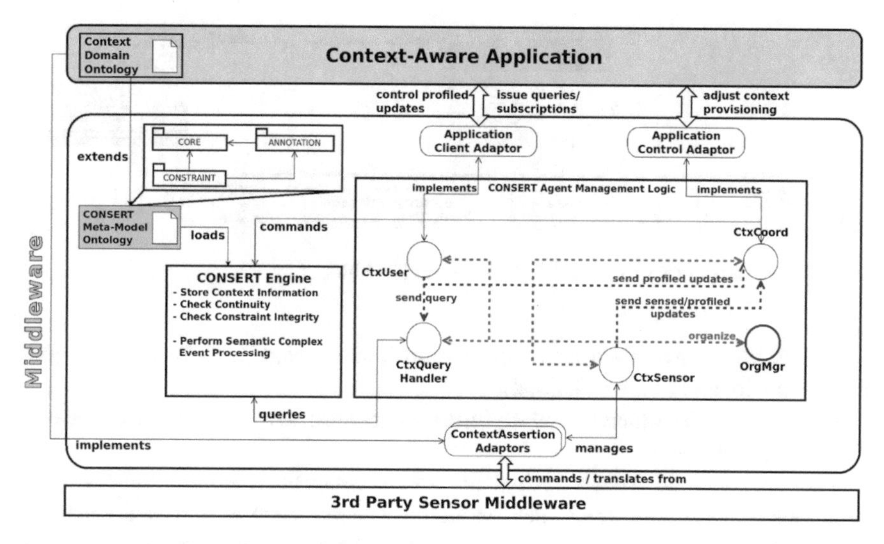

Fig. 5.3 CONSERT middleware block architecture overview

- CtxSensor agent: is responsible for managing interactions with sensors and with the CtxCoord agent of its own CMU, to handle provisioning commands (e.g. start/stop sending updates, change update rate).
- CtxCoord agent: is in charge of the management of the main life cycle of a CMU. It encapsulates a CONSERT Engine to control context reasoning and integrity.
- CtxQueryHandler agent: is in charge of the dissemination of context information. By default, it uses the local knowledge base to answer to queries. In case of more complex settings, involving multiple agents, it can participate in a decentralized federation protocol.
- CtxUser agent: is in charge of interfacing with the application. It exposes an Application Client Adaptor service interface which allows the application to launch queries and subscriptions, to send profiled context information (e.g. act as a sensor) or static entity descriptions.
- OrgMgr agent: is responsible for controlling the deployment of a CMU and of the entire middleware in cooperation with other OrgMgr agents: launch and control the states of CMM agents (started/stopped/uninstalled). It acts as a Yellow Pages agent, manages mobility aspects and maintains the overview of the Context Domain hierarchy in the decentralized deployment setting.

The agents communicate context information that is formatted according to the CONSERT Meta-Model, described in Fig. 5.4. The central work element is the *ContextAssertion*, which describes a certain action or state of affairs [e.g. *personLocated* (*John*, *kitchen*), *notificationSent* (*John*, *EarlyMorningWeightMeasurement*), *measurement* (*BloodPressure*, *John*, 120, 80)].

ContextAssertions can be predicates of any arity in which one or more *ContextEntities* (e.g. a Person, a Room, a Notification, numerical values) play a role. The

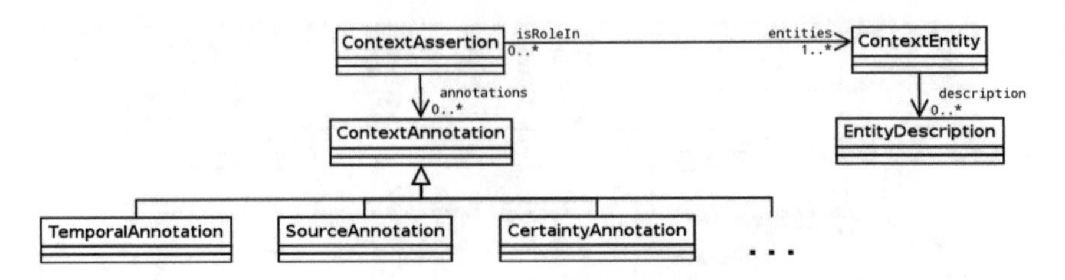

Fig. 5.4 Diagram of the CONSERT meta-model building blocks

ContextEntities can have further descriptive attributes, called *EntityDescriptions* (e.g. the ID and content of a notification).

ContextAssertions themselves are further characterized by meta-properties called *ContextAnnotations*.

The latter include aspects such as the *timestamp* or the *temporal validity* of the situations (ContextAssertions may be atomic events—such as a measurement, or they may have a duration in time—such as the presence in a room), the *source* of the information (e.g. which motion sensor was triggered, which health measurement device was used) or the *certainty* in the content of the event (e.g. motion sensors may have a degree of confidence in their reading depending on the distance to the user).

A complete description of the protocols that underlie communication between the CONSERT Middleware agents or the reasoning cycle of the CONSERT Engine is out of the scope of this paper. These have been covered in previous work [14, 16, 17].

However, in what follows, we show how CONSERT is usefully instantiated to manage health analysis and notification/reminder management within the CAMI System.

5.5.2 *Using the CONSERT Middleware in CAMI*

In Sect. 5.4 we explained that the CAMI architecture is conceived as a highly flexible composition of micro-services, each deployed in its own Docker container.

The instance of the CONSERT Middleware which handles health measurement analysis and notification management is no exception.

5.5.2.1 Modeling Context Information in CAMI

We mentioned that the Event Stream Manager is implemented as an instance of a RabbitMQ message broker instance, in which all measurement, home monitoring or user feedback events (e.g. acknowledging a reminder) are inserted.

Table 5.1 Example list of *ContextEntities* which have additional descriptions

ContextEntity	EntityDescription	
Person	hasID hasName hasProfileLang hasTimezone	String literal (URI) String literal String literal (enumeration) String literal
Notification	hasID hasType hasTitle hasContent	String literal (URI) String literal String literal String literal

Table 5.2 Example of *ContextAssertions* used in the CAMI system

ContextAssertion	Roles (ContextEntities)	ContextAnnotations
BPMeasurement	Person BPValues	Timestamp Source device Source gateway
WeightMeasurement	Person Numeric literal (weight value)	Timestamp Source device URI Source gateway URI
Motion	String literal (room name)	Timestamp Source sensor URI Source gateway URI
ExerciseStarted	Person String literal (exercise name)	Timestamp
ExerciseEnded	Person String literal (exercise name)	Timestamp
SendNotification	Person Notification	Timestamp
AckNotification	Person Notification	Timestamp

The *ContextEntities* which play a role in the assertion and necessary *ContextAnnotations* are indicated

To begin with, the relevant *ContextAssertions* and *ContextEntities* are modeled using the CONSERT Meta-Model. Tables 5.1 and 5.2 show a sample of EntityDescriptions, as well as modeled assertions and entities.

5.5.2.2 Deploying the CONSERT Middleware

In the CAMI DSS we currently use a single Context Management Unit which has the following composition:

- 2 CtxSensor agents: one that handles the health measurement information and one that handles the home monitoring and user interaction information.

- One CtxCoordinator and one CtxQueryHandler which manage a CONSERT Engine instance.
- One CtxUser instance which subscribes for all instances of created notifications/reminders which the system needs to send to a user.

Figure 5.5 shows how the CONSERT Middleware is instantiated in the CAMI DSS. The two mentioned CtxSensor agents are coupled via adapters to the corresponding channels in the Event Stream Manager. The adapters implement a RabbitMQ client which is subscribed to the required event queue (e.g. the Health Measurement Adapter subscribes to the Health channel). The adapter takes the measurement message, formatted according to the CAMI insertion API mentioned in Sect. 5.4, and translates it into *ContextAssertions* and *ContextEntities* of the form shown in Tables 5.1 and 5.2.

In the scenarios presented at the beginning of this section we explained that the output of the Health Analysis and Notification Management services are themselves notifications (e.g. alerts for abnormal health readings, reminders for taking health measurements). That is why the CAMI Application Adapter which links to the CtxUser instance instructs the agent to subscribe for instances the *send_notification ContextAssertion* which are derived by the CONSERT Engine.

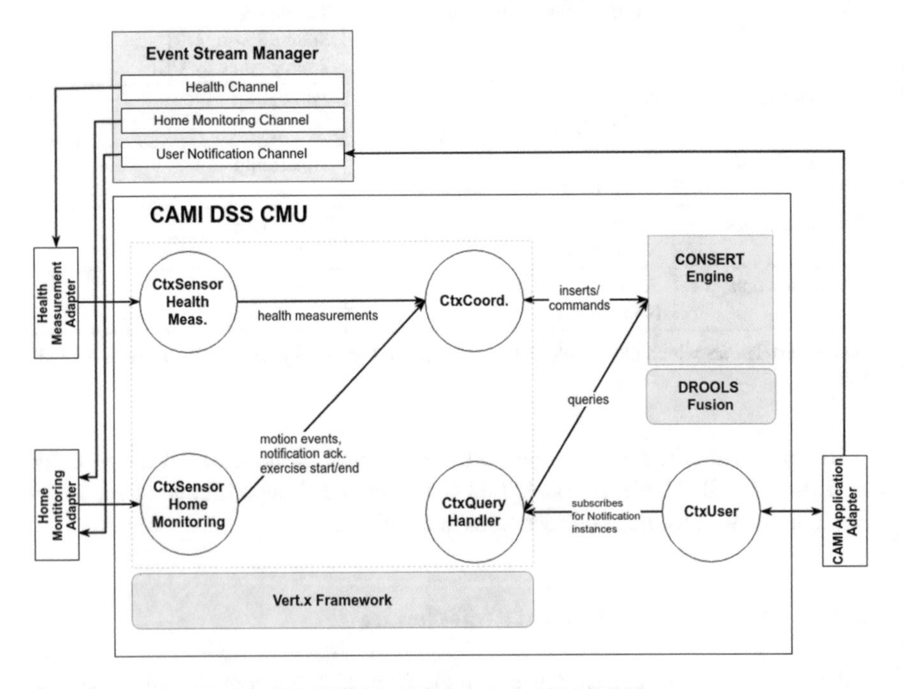

Fig. 5.5 Instantiation of the CONSERT middleware in the CAMI DSS

For each such assertion, the adapter converts it into a JSON payload and publishes it on the User Notification Channel of the Event Stream Manager. From there it is picked up by a CAMI Cloud Service which sends notifications to end user smartphones via the PushBots API [18].

5.5.2.3 Inferring Context Information

Inferences in the CONSERT Middleware are performed by the CONSERT Engine. The latter can be succinctly described as a Semantic Complex Event Processing engine that can implicitly handle temporal validity extensions of derived context situations.

Unlike traditional CEP engines, which execute various filtering or aggregation operations on a stream of data, the CONSERT Engine offers a specific information flow that addresses the reasoning challenges commonly found in AmI applications. It specifically implements an explicit temporal validity extension mechanism, as well as a constraint resolution one. Details of these methods are given in [16, 17].

The engine is built on top of the DROOLS Fusion framework, which give it benefits such as: rule saliency, time-window and count-based aggregation, use of the *not* operator to express absence of events over time.

In the following we discuss about a selected number of rules that implement the functionality described in the scenarios from the beginning of the section.

To begin with, we have mentioned that the CAMI System can register events triggered by motion sensors such as the Fibaro Motion Sensor [19]. However, in the presented scenario, we want to send notifications for early morning weight measurements when the user is first detected in the bathroom. Therefore, we want to trigger the rule when we are sure that the user is in the bathroom and is staying there for a period of time (e.g. to do morning toileting).

Listing 5.1 displays an inference rule that detects when a person is staying in a room. The rule states that if a person is in the bathroom and there is no other

```
rule "Remain in the Bathroom"
when
        $loc : PersonLocation(p: person, room : loc == "Bathroom",
                locAnn : annotations)
        not( exists PersonLocation(person == p, loc != room,
                this annOverlappedBy[0s, 5s] $loc || $loc annIncludes this))
        not( exists Motion(status == "ON", this annHappensAfter[0s, 5s] $loc))
then
        long ts = eventTracker.getCurrentTime();
        DefaultAnnotationData ann = new DefaultAnnotationData(ts);
        PersonLocation sameLoc = new PersonLocation("Bathroom", ann);
        eventTracker.insertAtomicEvent(sameLoc);
end
```

Listing 5.1 Context inference rule that activates when a person is in the kitchen and stays there

```
rule "Early Morning Weight Measurement"
when
        $loc : PersonLocation(p: person, room : loc == "Bathroom",
                locAnn : annotations, locAnn.startTime > 6AM, locAnn.endTime < 11AM,
                locAnn.duration >= 1m)
        not( exists PersonLocation(person == p, loc == room,
                this annHappensBefore[0s, 5h] $loc))
        not( exists WeightMeasurement(person == p, this annHappensAfter[0s, 10m] $loc)
                || $loc annIncludes this)
        not( exists SendNotification(person == p, notif : notification,
                notif.type == "weight_measurement")))
then
        // send the reminder for early morning weight measurement
        long ts = eventTracker.getCurrentTime();
        DefaultAnnotationData ann = new DefaultAnnotationData(ts);
        Notification n = new Notification(p, "weight_measurement");
        SendNotification sendNotif = new SendNotification(p, n, ann);
        eventTracker.insertAtomicEvent(sendNotif);
end
```

Listing 5.2 Rule for sending an early morning weight measurement notification, after first visit to the bathroom is detected

overlapping location for that same person and no other Motion sensor is triggered in the next 5 s, then the current person location can be extended.

Notice the use of the *not* operator, which in this case allows us to condition on the absence of other events during a period of time, before or after an existing event.

If we can determine when a person is in the bathroom, the rule in Listing 5.2 shows how to send a notification for an early morning weight measurement. The first two *PersonLocation* conditions ensure that this is the first time in the morning, that the user visits the bathroom, a sign he has woken up. The next condition waits for a weight measurement from the user for at most 10 min, considering the user might remember by himself to take the measurement. The last condition verifies that the system hasn't already sent such a notification. Lastly, Listing 5.3 shows how the CONSERT Engine checks for abnormal decreases in two consecutive weight measurements.

```
rule "Abnormal Weight Decrease"
when
        $meas1 : WeightMeasurement(p: person, v1 : value, ann1 : annotations)
        $meas2 : WeightMeasurement(person == p, v2 : value, v2 - v1 > 2,
                this annHappensBefore $meas1))
        not( exists WeightMeasurement(person == p, this annHappensAfter $meas2,
                this annHappensBefore $meas1))
then
        // send a notification for abnormal weight decrease
        long ts = eventTracker.getCurrentTime();
        DefaultAnnotationData ann = new DefaultAnnotationData(ts);
        Notification n = new Notification(p, "weight_decrease");
        SendNotification sendNotif = new SendNotification(p, n, ann);
        eventTracker.insertAtomicEvent(sendNotif);
end
```

Listing 5.3 Rule for sending a notification following abnormal decrease in two consecutive weight measurements

The first two conditions identify the weight measurements for a person, while the third one ensures they are consecutive (e.g. no other measurement is in between them).

The overall benefit of using a CONSERT Middleware instance to handle part of the CAMI DSS responsibilities is the level of control this brings to the development and maintenance process. The CONSERT Meta Model is well suited for the information representation requirements of the CAMI DSS and the declarative nature of the rules in the CONSERT Engine ensures a good level of comprehensibility for the intended functionality of the system, such that debugging any errors becomes easier.

Figure 5.5 shows that the *SendNotification* ContextAssertions to which the CtxUser agent is subscribed are subsequently inserted back in the EventStream-Manager to be picked up and delivered to end-user smartphone devices by the CAMI push notifications forwarding service. The way in which the end-user can visualize these notifications and other health related information is covered in the next section.

5.6 Multimodal Interaction with the User

Since elderly people are the main targeted users of the CAMI system, the interface was designed by respecting their different requirements such as the displayed elements size and the ways of interactions. To facilitate the interaction with the system from a part and to make the interactions more natural from the other part, the interface supports beside the traditional inputs, speech and gesture commands. The multimodal interface integrates an avatar that interacts with the user through speech and text, it can be activated or deactivated by the user easily at any time. As the targeted users are from different countries, the users come with various backgrounds, culture, knowledge, languages and with different preferences; a big challenge is to satisfy all the users. Therefore, the interface has adaptive capabilities, it works cross platforms and it is independent of the device or the operating system. It adapts itself to each device screen size but also to each user profile, preferences, emotional state and to the system configuration. The interface is multilanguage, it supports different languages of the CAMI consortium (English, French, Romanian, Swedish, Danish, Polish and Italian), with the exception of the voice interaction module that supports in the actual stage, only French, English and Romanian languages; with the possibility of adding, easily, additional languages. The interface illustrated in Fig. 5.6, is composed of 4 modules: voice interaction, gesture, emotion and visual modules. The user can also interact with the system with the help of a robot. As the system should stay accessible in the case of a lose or absence of internet connection, the interface has two modes of working: one that is used normally and depends on the internet connectivity (online/normal mode) and the other that doesn't depend on the internet connectivity and that is used in case of connection lost/absence (offline/limited mode).

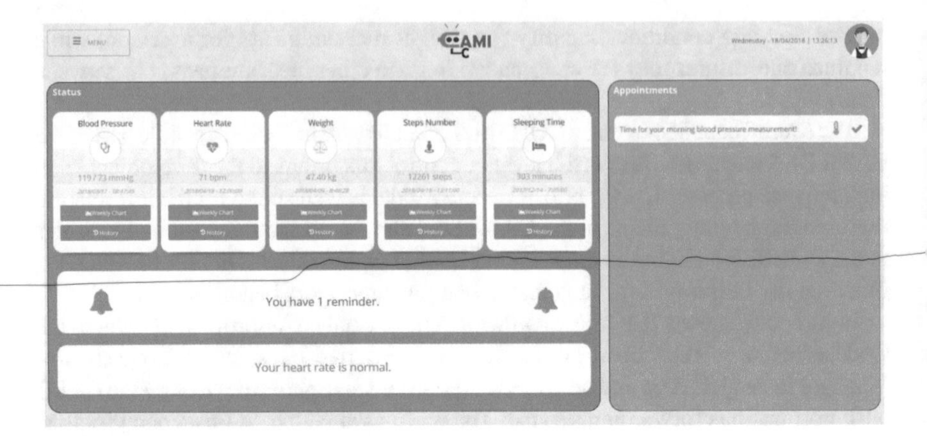

Fig. 5.6 Interface preview: main page

5.6.1 *Multimodal Interface Implementation*

As we mentioned earlier, the multimodal interface is composed from multiple modules. We first discuss the voice interaction module, followed by the gesture module, the emotion module and the visual module. At the end, we describe shortly the possibility to interact with the CAMI system through a robot.

The voice interaction module of the multimodal interface allows oral (speech and non-speech) interactions between the user and the system. It allows the user to interact with the system through speech commands: navigate into the system pages, get information or order actions to be executed. After analyzing the existing solutions in the field, we developed a voice interaction module that supports English, French and Romanian languages. It is composed of five main components: Audio Preprocessing, Automatic Speech Recognition (ASR), Natural Language Understanding (NLU), Dialog Management (DM), and Text to Speech synthesis (TTS).

Audio Preprocessing: The Audio Preprocessing ensures the best results for the ASR by setting-up different parameters that are useful for the recognition, based on the information that are extracted from the profile of the user and the system settings, such as the speech language, the frequency of the audio, the number of channels of the audio, the ASR solution that should be used, and as well as other useful properties for the voice module.

ASR component: Briefly, ASR is the technology that allows a computer to identify the words that a person speaks into a microphone or telephone and convert them to written text. In 1994, Stuckless [20] defined ASR as an independent, computer-driven transcription of spoken language into readable text in real time. For the normal mode of working, we used the speech-to-text service provided by Microsoft Azure as our ASR solution for the English and French languages. Meanwhile, we used the speech-to-text service provided by Google Cloud as our ASR solution for the Romanian language. For the limited mode of working, we used the Microsoft Windows Speech

Technology as our ASR solution. In the limited mode of working, the numbers of commands recognized by the system is reduced to the set of basic ones determined in the system settings and the Romanian language recognition is disabled.

NLU component: The NLU component converts the words of the user's utterance received as input from the ASR to a machine-reading representation. Its aim is to extract semantic meaning from the received text. It is also responsible to correct any errors made by the ASR. For the normal mode of working, we used the Language Understanding Intelligent Service (LUIS) from Microsoft as our NLU solution for the English and French languages while we used the wit.ai API. For the limited mode of working, we used RASA NLU provided by Rasa Technologies GmbH as the NLU solution.

DM component: The DM component is responsible for the state and flow of the conversation. It receives as input the machine-reading representation of the user's utterance generated by the NLU. Its output is a semantic representation of a list of instructions for the dialog system. The list determines which should be the system's answer in response to the user's processed input. For the normal mode of working, we used the Azure Bot Service provided by Microsoft as the DM solution, while we used, for the offline mode of working, the Bot Builder SDK provided by Microsoft to build the DM solution.

TTS component: The TTS component artificially produces any generated "normal language text" output into human speech that is heard over the speakers of the system. For the normal mode of working, we used the text-to-speech service provided by Microsoft Azure as our TTS solution for the English and French languages, while we used the ResponsiveVoice.JS API as our TTS solution for the Romanian language. For the limited mode of working, we used the Microsoft Windows Speech Technology as our TTS solution.

The gesture module of the multimodal interface allows gesture interactions between the user and the system. Using a Kinect V2 sensor (or any other device that have capability of motion acquisition) and mathematical algorithms, gesture recognition allows machines to interpret and understand human body language which represents inputs for the system.

The emotion module of the multimodal interface allows the system to track the emotional status of the user using the face features. The system uses a Kinect V2 sensor and mathematical algorithms to understand the emotional state of the user by interpreting the facial expressions of the user. The user's facial emotions are recognized based on features extracted from the user's face: reference points from face and intensities of the action units, as described in [21]. The reference points are obtained using Microsoft.Kinect.Face and Microsoft.Kinect [22]. Using the Microsoft.Kinect.Face 1000 reference points from the user 's face are obtained. Only 51 are selected: 5 points for each eye brow, 6 points for each eye, 11 points for nose and 20 points for mouth. These 51 points are mapped to action units associated to the basic emotions. Support vector regression networks are using in order to recognize 7 basic emotions: happiness, sadness, surprise, fear, anger, disgust and contempt.

The visual module of the multimodal interface allows the system to provide information to the user through different screens that are connected to the system. It adapts itself to the screen size and to the device orientation. The design of the visual module is optimized for touch screens. The main page of the interface displays the last health measurements results, the daily number of steps and the sleep duration of the user for the current day. It displays also the daily appointments of the user, the number of reminders and some smart notifications regarding the users' health status and activity. The smart notifications colors as well as the colors of the icons of the measurements depends on its value. If the value of the measurement is within the normal range, the colors of its icon and its related notification is green. The colors are replaced by yellow if some small attention is needed and by red if serious attention is needed. Those colors are the default colors used by the interface, however the user can change those colors from the setting page. The normal range of values for each health measurement is defined according to a medical standard. However, the range of values can be easily adjusted for each user by the users' caregiver. The menu of the interface allows the user to navigate smoothly between the different pages of the interfaces: health history, health weekly chart, appointments, settings, etc. The interface allows the user to visualize weekly charts that illustrate the users' health measurements results, the daily number of steps and the sleep duration for the previous seven days. The user can also request to visualize his/her data for any period that he/she desire. The results of the request can be visualized in a form of a list or a chart and the user can easily switch between those two modes of visualization. In the limited mode of working, the interface displays an alert to inform the user about the internet connectivity status and it displays the last received measurements from the cloud.

In case of a multimodal interface, the multiple types of input can lead to some problems such as contradictory commands received by the system on the same time, through different input channels. To face this problem the interface integrates algorithms that analyze and choose what command should be considered. This choice is affected by the degree of confidence of the command recognition but also by different parameters extracted from the user profile.

- Each user profile contains information about the user health status, specific problems or factors that can affect the user's behaviour, previous interactions between the system and the user that had contradictory inputs, and as well other information.
- For each input modalities, the parameters that can be related to this modality or can affect it, are extracted and analyzed then the system gives a score to the input modality (the score can be positive or negative).
- The scores obtained from the different parameters are added to the degree of confidence of the corresponding detection.
- If the difference between the score is with a major difference for an input modality, the system considers the input from this modality and ignores the other ones, at the end of this interaction the user has the option to give a feedback regarding the chosen input. If the difference is in favor of an input but it's still not a major one, the system asks the user if he/she meant that. If the difference is small or if

there is no difference at all, the system asks the user to repeat the command again. At the end of each interaction that contain contradictory commands, the different elements of the interaction are stored in the user profile including the right input modality.

The interaction with the robot is a novel mode of interfacing with a health monitoring system, introduced in CAMI. Pepper is a humanoid friendly looking robot [23] which is used in the CAMI project. The robot has capabilities which allow it to identify, track and follow people, carry out simple dialogs and display information on the integrated tablet. In CAMI, the considered interaction scenarios involve asking Pepper to do the following:

- Display the current health status of a person based on voice commands.
- Look for, navigate to and identify an elderly person which has unacknowledged important reminders (e.g. to take some medication or carry out a health measurement).

The full details of the interaction scenarios and the technical integration specification are out of the scope of this paper, but they can be inspected in a related paper [24].

5.6.2 *Interface Evaluation*

The design and implementation of the different modules of the multimodal interface have been tested in several stages. Older implementations of the voice module have been tested in [25–27], the obtained results were satisfying for the ASR and TTS modules in both mode of working (offline and online), however moving to a cloud-based ASR and TTS services allows us to reduce the needed computation resources. The gesture module has been tested with four gestures, as illustrated in [26] and the obtained results were also satisfying: toward right (96.25% accuracy detection), toward left (96.25%), circle (58.75%) and square (72.50%). We extended the set of emotions illustrated in [28]. In this case we consider 7 basic emotions: happiness, sadness, surprise, fear, anger, disgust and contempt. The obtained results were as well satisfying: happiness (98.20% accuracy detection), sadness (89.30% accuracy), surprise (98.60% accuracy), fear (84% accuracy), anger (82% accuracy), disgust (89.50% accuracy), contempt (60% accuracy). The results and the feedback obtained from the tests helped us to improve the interface.

The visual module of the multimodal interface has been evaluated by 38 elderly people (20 male and 18 female) aged between 57 and 63 years old, on different devices: desktop computers, laptops, tablets (iOS, Android and Windows) and mobiles (iOS, Android and Windows). The results showed that the interface was fully responsive and worked across the different devices and platforms that has been tested on, as illustrated in Fig. 5.7.

The previous users have also tested the new implementation of the voice interaction module by performing an extended set of commands. The test has been done on

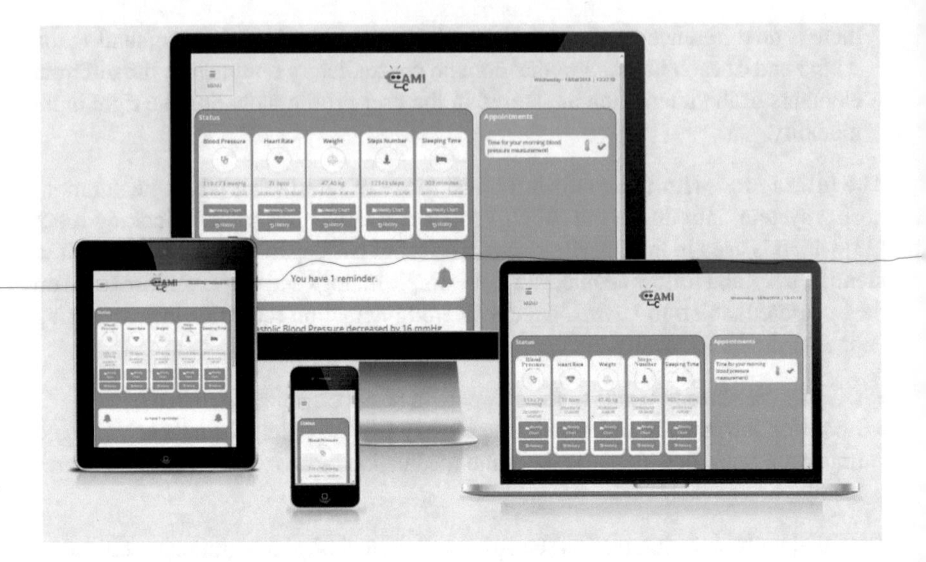

Fig. 5.7 The interface is fully responsive

a set of 2280 interactions. Each user executed 60 interactions with the system: 20 interactions in the Romanian language and 20 interactions in English or French (half by half) language but twice, once using the online mode and once using the offline mode. The tests have been done at the user's home using a Plantronics Voyager 5200 UC Microphone attached to an HP ZBook 15 G3 system (Core i7 2.60 Ghz, 8 GB RAM, integrated stereo speakers) which has Windows 10–64 bits as operating system. Table 5.3 illustrates some results of the ASR test while Table 5.4 illustrates some results of the TTS test.

Comparing with our previous work, the voice interaction module of the interface supports more commands in his both modes of working and the online mode works completely within the cloud. Moreover, the multimodal interface has additional adaptive capabilities: it can adapt itself to the configuration of the system and to more parameters of the user profile (e.g. most used commands, feedback type preferences, health condition, etc.).

5.7 Stimulating User's Physical Activity

The stimulation of the user's physical activity aims to (1) improve the user's lifestyle by providing regular physical activity that is consistent with his/her medical condition; (2) monitor the user's activity and provide motivational feedback; provide a personalized exercise program that can be adapted by his medical parameters, depending on patient progress. The application is developed using the Unity 3D engine and it is based on two avatars: the training avatar and the avatar of the user—the application

Table 5.3 Some ASR test results

Language	Users input	Average recognition percentage (%)	
		Online mode	Offline mode
English	Display my blood pressure	97.51	86.17
	How much have I walked today	96.91	85.56
	How will the weather be today	98.82	89.16
	Show my calendar	99.06	89.95
	What is my health status	97.35	86.64
French	Afficher ma tension artérielle	93.69	85.58
	Combien j'ai marché aujourd'hui	92.12	79.96
	Comment sera le temps aujourd'hui	93.95	86.02
	Afficher mon calendrier	96.71	85.98
	Quel est mon état de santé	94.52	84.27
Romanian	Arată-mi tensiunea arterial	91.15	N/A
	Cât am mers astăzi	94.11	N/A
	Cum va fi vremea astăzi	94.65	N/A
	Afișează-mi calendarul	95.43	N/A
	Care este starea mea de sănătate	94.89	N/A

Table 5.4 Some TTS test results

Language	System spoken output	User satisfaction (average score)	
		Online mode	Offline mode
English	You have a new notification	9.75	9.50
	Here is your weekly calendar	10.00	9.75
	You did well	10.00	10.00
French	Vous avez une nouvelle notification	9.75	9.50
	Votre calendrier hebdomadaire est affiché	9.25	8.25
	Tu as bien fait	10.00	9.75
Romanian	Aveți o notificare nouă	8.75	N/A
	Acestea este calendarul săptămânal	7.75	N/A
	Te-ai descurcat bine	7.50	N/A

is described in [21, 29]. The training avatar performs different physical exercises and the user must reproduce his movements. The user's movements are captured using the Kinect v2 sensor. The stimulating user's physical activity component is composed of: (i) selecting the type of the exercise based on the current values of the monitored medical parameters, (ii) performing the exercise and obtaining the final score, (iii) feedback to the user and (iv) selecting another exercise.

Selection for the type and the intensity of the exercise is based on the monitored medical parameters and the physician recommendations. Each user has associated a list of possible exercises. The list is created by a physician based on the user's medical profile. From this list one exercise is selected associated with its intensity based on current values of the user's heart rate and pulse.

In order to compute the score associated to the performed exercise by the user, the movements are described using the skeleton joints provided by the Kinect V2 sensor. The sensor provides 25 joints for a user at approximately 30 frames per second. We use only 20 joints are used. For each joint we compute a 3D rotation using quaternions relative to the parent bone. These joints are grouped in a hierarchy, and the rotation value for each joint is calculated in the 3D coordinate space relative to the parent bone. Figure 5.8 describes the hierarchy of joints used to detect the user's posture.

In order to compare the trainer's and user's movements, the similarity between the set of quaternion values for each joint computed per frame is compared using Dynamic Time Warping (DTW) algorithm. DTW computes the similarity between two series by calculating the minimum distance between them. It is a much robust way of calculating the degree of similarity between two-time data sequences, identifying similar forms even if they are not aligned along the time axis. A screenshot of the game is given in Fig. 5.9.

The application aims to personalize the exercises according to the medical condition of each user. Thus, we associate weights with each joint in order to reflect the user physical state. Each weight has a value in [0, 1] and is set by a physician.

In order to compute similarities between the movements performed by the user and by the trainer we use different distance metrics: inner product, Euclidean distance, square Euclidean distance and Manhattan distance as given in [30].

We consider different execution for the same exercise: (1) the user performs the exercise similar to the reference, (2) the user performs the exercise slower than the trainer, (3) the user performs the exercise partially incorrect and (4) the exercise is executed more slowly and completely differently from the exercise performed by the trainer. To evaluate the accuracy of the generated score, the application was tested on an extended group of users—in this case we consider 20 users, each executing the

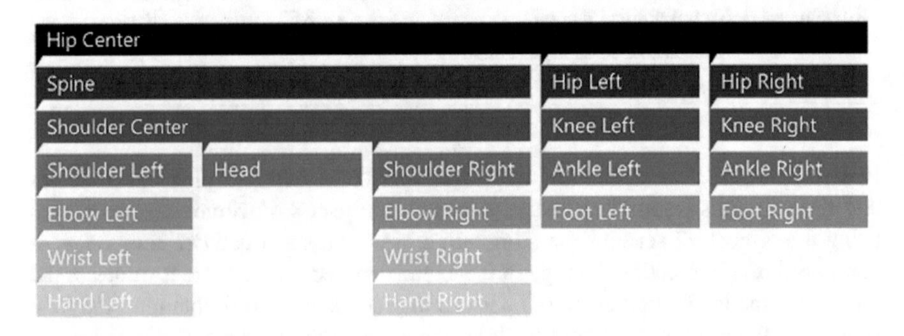

Fig. 5.8 The hierarchy of joints for the human skeleton

Fig. 5.9 Screenshot of the game

same reference exercises. We tested the following exercises: hip extension, squats, lateral lunge, quadriceps stretch, lateral stretch and arm stretches. Using the square Euclidean distance and the inner distance we can differentiate between the four cases that we considered, as given in Fig. 5.10.

We integrated a feedback component in order to provide feedback to the user regarding his evaluation. The feedback is given through the multimodal interface.

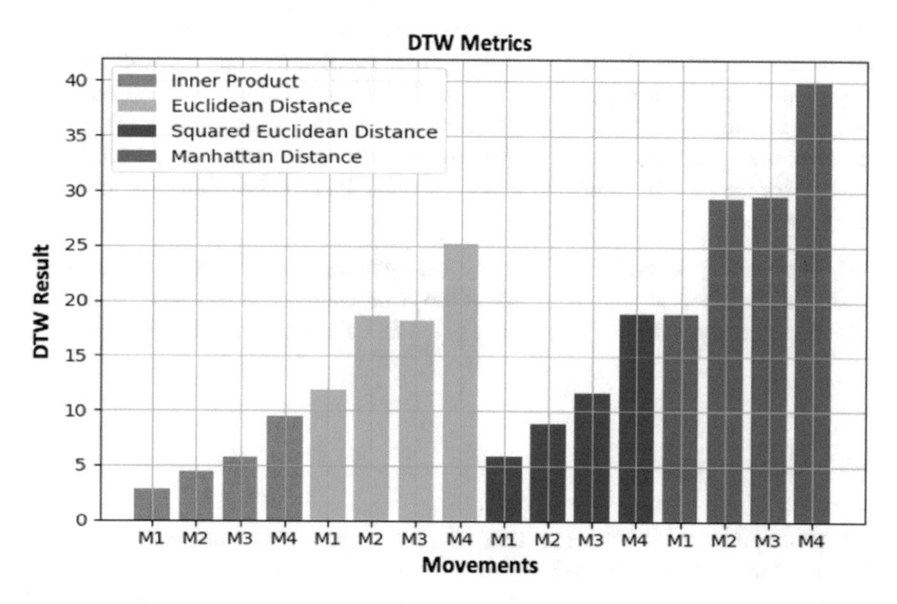

Fig. 5.10 Different metric distance

At the end of the session, an automated review that result of the movements analysis appears on the screen. It includes an avatar that replay the last exercise performed by the user. The review includes the mistakes committed by the user during the session. A score and some suggestions that improve user's performance are also included in the review.

During the exercise if the value of the user's pulse (obtained from the Fitbit sensor) is increasing, the user will be notified by the CAMI system in order to stop the physical exercises and rest for a while.

After performing an exercise, the user can be advised by the system to perform another exercise. The recommendation of continuing making physical exercises is based on the following health parameters monitored by the CAMI system: number of steps and the user's pulse monitored during the exercise. Also, the type of exercise will be recommended to the user—the type of the next exercise will be selected based on the current pulse of the user and also based on the predominant user' facial emotions during the last exercise. Using the emotion module from the *Multimodal Interaction with the User*, we recognize the user's face emotions. We create two groups of emotions: (1) *group 1*: happiness, surprise; and (2) *group 2*: sadness, fear, anger, disgust and contempt. We create 2 rules for selecting the exercise type:

- If the current detected emotion is part of the group 1, we maintain the exercise.
- If the current detected emotion is part of the group 2, the user is not happy and the system we'll change the type of the exercise in order to keep the user making physical activity.

Finally, the user receives an evaluation of its performance, as: *excellent, good, bad*. The performance of the user is evaluated using an adaptive neuro-fuzzy inference system composed of a neural network and a fuzzy logic system [31]. The adaptive neuro-fuzzy inference system uses learning paradigms as a neural network by mapping modules of the fuzzy logic system to the layers of the neural network. As in [31], we consider two-stage fuzzy inference system that evaluates the user's performance two adaptive neuro-fuzzy inference: one for the trajectories of the users' joins and one for the time in which the user performed the exercise. Rules for the fuzzy inference are:

- if the trajectory is similar and speed is just right, then the evaluation is *excellent*;
- if the trajectory is similar and speed is too slow, then the evaluation is *good*;
- if the trajectory is dissimilar or speed is too fast, then the evaluation is *bad*.

5.8 Conclusions

This paper presents the CAMI system, which fully integrates a range of AAL solutions by offering health, home care, social and other services. Since elderly people are the main target users of the system, CAMI services are flexible, scalable, and individualized to satisfy each user's need.

The system interacts with the users through a multimodal interface that allows natural means of interaction and is designed based on the elderly users' requirements. The CAMI system monitors the vital signs of the users, stores and analyze them in the cloud, recommends physical exercises to users, depending on their medical status and adapts the performed physical exercises to their current performances.

After analyzing some previous AAL solutions, we gave details about the range of functionalities that the CAMI system provides. We illustrated the architecture of the system and we detailed the functionality of the framework that underlies the intelligent health analysis and notification management service in the CAMI System. We described the multimodal interface of the system and we detailed the physical exercise monitoring module which enhance the mobility of the user. Finally, we illustrated some results of the multimodal interface test.

As future work, extensive field trials are planned to gather feedback from the users regarding the implementation of the CAMI functionalities. In addition, we are planning to extend the supported languages of the voice interaction module, as well as the customizable features of the interface.

Acknowledgements This work was supported by the following two programmes: Active and Assisted Living program through a grant of the Romanian National Authority for Scientific Research and Innovation, CCCDI—UEFISCDI, CAMI—"The Artificially intelligent ecosystem for self-management and sustainable quality of life in AAL", project number AAL-2014-1-087 and National Research Grant PN-III-P2-2.1-PED-2016-1753.

References

1. Cayton, H.: The flat-pack patient? Creating health together. Patient Educ. Couns. J. **62**, 288–290 (2006). https://doi.org/10.1016/j.pec.2006.06.016
2. Darwish, M., Senn, E., Lohr, C., Kermarrec, Y.: A comparison between ambient assisted living systems. In: 12th International Conference on Smart Homes and Health Telematics (ICOS), pp. 231–237. Denver, Colorado, USA, https://doi.org/10.1007/978-3-319-14424-5_26 (2014)
3. Tazari, M.R., Furfari, F., Ramos, J.P.L., Ferro, E.: The persona service platform for AAL spaces. In: Handbook of Ambient Intelligence and Smart Environments, pp. 1171–1199. Springer (2010)
4. Lamprinakos, G., Kosmatos, E., Kaklamani, D., Venieris, I.: An integrated architecture for remote healthcare monitoring. In: Proceedings of the 14th Panhellenic Conference on Informatics, pp. 12–15. Tripoli, Greece, https://doi.org/10.1109/pci.2010.20 (2010)
5. Active and Assisted Living Program: healthy@work, retrieved from the project page on the official AAL website. http://www.aal-europe.eu/projects/healthywork/. Visited on: 21 June 2018
6. Planinc, R., Hödlmoser, M., Kampel, M.: Enhancing the wellbeing at the workplace. In: The 7th International Conference on eHealth, Telemedicine, and Social Medicine (eTelemed), pp. 213–216. Lisbon, Portugal (2015)
7. Azevedo, C., Chesta, C., Coelho, J., Dimola, D., Duarte, C., Manca, M., Nordvik, J., Paterno, F., Sanders, A., Santoro, C.: Towards a platform for persuading older adults to adopt healthy behaviors. In: Orji, R., Reisinger, M., Busch, M., Dijkstra, A., Kaptein, M., Mattheiss, E. (eds.) Proceedings of the Second International Workshop on Personalization in Persuasive Technology, pp. 50–56. Amsterdam, Netherlands (2017)

8. Giannoglou, V., Smagas, K., Valari, E., Stylianidis, E.: Elders-up! an adaptive system for enabling knowledge transfer from senior adults to small companies. In: 22nd International Conference on Virtual System & Multimedia (VSMM), pp. 17–23, Kuala Lumpur, Malaysia. https://doi.org/10.1109/vsmm.2016.7863163 (2016)
9. http://www.andonline.com/medical, visited on 29 June 2018
10. https://www.openhab.org, visited on 29 June 2018
11. http://www.docker.io, visited on 29 June 2018
12. https://www.rabbitmq.com, visited on 29 June 2018
13. https://github.com/cami-project/cami-project/blob/master/insertion/cami-insertion-api.yml, visited on 29 June 2018
14. Sorici, A., Picard, G., Boissier, O., Florea, A.M.: Multi-agent based flexible deployment of context management in ambient intelligence applications. In: International Conference on Practical Applications of Agents and Multi-Agent Systems, pp. 225–239. Springer, Cham (2015)
15. American Heart Association. Understanding blood pressure readings, June 2018. Retrieved from http://www.heart.org/HEARTORG/Conditions/HighBloodPressure/KnowYourNumbers/Understanding-Blood-Pressure-Readings_UCM_301764_Article.jsp, visited on 29 June 2018
16. Sorici, A., Picard, G., Boissier, O., Zimmermann, A., Florea, A.M.: CONSERT: applying semantic web technologies to context modeling in ambient intelligence. Comput. Electr. Eng. **44**, 280–306 (2015)
17. Trăscău, M., Sorici, A., Florea, A.M.: Detecting activities of daily living using the CONSERT engine. In: Novais, P., et al. (eds.) Ambient Intelligence—Software and Applications, 9th International Symposium on Ambient Intelligence (ISAmI2018), Advances in Intelligent Systems and Computing, vol. 806, pp. 94–102. Springer, Cham (2018)
18. https://pushbots.com, visited on 29 June 2018
19. https://www.fibaro.com/en/products/motion-sensor, visited on 29 June 2018
20. Stuckless, R.: Developments in real-time speech-to-text communication for people with impaired hearing. In: Ross, M. (ed.) Communication access for people with hearing loss, pp. 197–226. York Press, Baltimore, MD (1994)
21. Mocanu, I., Schipor, O.A.: A serious game for improving elderly mobility based on user emotional state. In: 13th eLearning and Software for Education Conference, vol. 2, pp. 487–494. Bucharest, Romania (2017)
22. Kinect for Windows SDK 2.0, https://www.microsoft.com/en-us/download/details.aspx?id=44561, visited on Nov 2017
23. Softbank robotics. Who is pepper? Retrieved from www.softbankrobotics.com/emea/en/robots/pepper, visited on 29 June 2018
24. Ghiță, Ș.A., Barbu, M.S., Gavril, A.F., Trăscău, M., Sorici, A., Florea, A.M.: User detection, tracking and recognition in robot assistive care scenarios. In: Giuliani, M., Assaf, T., Giannaccini, M. (eds.) Towards Autonomous Robotic Systems (TAROS2018), Lecture Notes in Computer Science, vol. 10965, pp. 271–283. Springer, Cham (2018)
25. Awada, I.A., Codreanu, A., Mocanu, I., Florea, A.M., Apostu, M.: An adaptive multimodal interface to improve elderly people's rehabilitation exercises. In: 13th eLearning and Software for Education Conference, vol. 2, pp. 41–47. Bucharest, Romania (2017)
26. Awada, I.A., Mocanu, I., Florea, A.M., Cramariuc, B.: Multimodal interface for elderly people. In: 21st International Conference on Control Systems and Computer Science (CSCS), pp. 536–541. IEEE, Bucharest, Romania (2017)
27. Awada, I.A., Mocanu, I., Florea, A.M.: Exploiting multimodal interfaces in eLearning systems. In: 14th eLearning and Software for Education Conference, vol. 2, pp. 174–181. Bucharest, Romania (2018)
28. Awada, I.A., Mocanu, I., Rusu, L., Arba, R., Florea, A.M., Cramariuc, B.: Enhancing the physical activity of older adults based on user profiles. In: 16th RoEduNet Conference: Networking in Education and Research (RoEduNet), pp. 120–125. IEEE, Targu Mures, Romania. ISSN: 2068-1038. https://doi.org/10.1109/roedunet.2017.8123749 (2017)

29. Mocanu, I., Rusu, L., Arba, R., Marian, C.: A kinect based adaptive exergame. In: 12th International Conference on Intelligent Computer Communication and Processing, pp. 117–124. Cluj-Napoca, Romania, https://doi.org/10.1109/iccp.2016.7737132 (2016)
30. Mocanu, I., Caciula, R., Gherman, L.: Improving physical activity through exergames. In: 14th eLearning and Software for Education Conference, vol. 2, pp. 225–232. Bucharest, Romania (2018)
31. Su, C.J., Chiang, C.Y., Huang, J.Y.: Kinect-enabled home-based rehabilitation system using dynamic time warping and fuzzy logic. Appl. Soft Comput **22**, 652–666 (2014). https://doi.org/10.1016/j.asoc.2014.04.020

Chapter 6
Machine Learning Based Assistive Speech Technology for People with Neurological Disorders

Shanmuganathan Chandrakala

Abstract With the tremendous improvements of automatic speech recognition systems worldwide, efficient ways of recognizing dysarthric speech has emerged as a practical challenge. Recognizing the impaired speech with poor articulation, missing consonants, and so forth is one of the foremost requirements in research for speech domain. Given an unknown dysarthric (partial) speech utterance, the problem is to recognize the speech content. I first review and analyze the different approaches such as generative, discriminative, hybrid model based approaches and unsupervised approaches for dysarthric speech recognition (DSR). Next, I present a framework in which effective representations are formed using generative model-driven features for dysarthric speech recognition task. The performance of the proposed method is examined to recognize the isolated utterances from the UA-Speech database. The recognition accuracy of the proposed approach is better than the conventional hidden Markov model-based approach.

Keywords Dysarthric speech recognition (DSR) · Mel frequency cepstral coefficients (MFCC) · Generative model-driven features · Hidden Markov models · Gaussian mixture models · Likelihood embedding-support vector machine · Transition embedding-support vector machine

6.1 Introduction

Neurological disorders are the diseases that affect the central nervous systems and the brain. The physical symptoms of neurological problems may include muscle weakness, seizures, poor cognitive abilities, difficulty in writing and reading, partial or complete paralysis, tiredness, slurred speech and so forth. Many people around the world are affected by different types of neurological disorders. Few types of disorders are Alzheimer's disease, epilepsy, multiple sclerosis, Parkinson's disease and

S. Chandrakala (✉)
Intelligent Systems Group, School of Computing, SASTRA University, Thanjavur, Tamil Nadu, India
e-mail: chandrakala@cse.sastra.edu

H. Costin et al. (eds.), *Recent Advances in Intelligent Assistive Technologies: Paradigms and Applications*, Intelligent Systems Reference Library 170,
https://doi.org/10.1007/978-3-030-30817-9_6

migraines. Parkinson's disease (PD) causes motor and non-motor symptoms. One such motor symptom is the speech difficulty found in patients suffering from PD. The voice may become breathy, hoarse and make the speech unclear. Speech may be slurred or expressed rapidly. Imprecise articulation, slow speech, stuttering, tremor, too nasal, impaired stress or speech rhythm is commonly found voice and speech problems with people suffering from Parkinson's disease and multiple sclerosis. People suffering from Alzheimer's disease (AD) [40] may find overlaps in speech comprehension, production of speech, and memory functions. The language measures include fluency in finding words, semantics and articulation. The verbal fluency may be influenced by short-time memory. The speech problems occurring to such neurological orders are called dysarthria [10]. This article reviews the pattern recognition literature to automatic speech recognition (ASR) systems to recognize dysarthric speech. Automatic speech recognition is a process in which a machine recognizes the speech and translates into text. Speech recognition systems are advantageous in the way that it avoids/minimizes the usage of keyboard and mouse. ASRs have been applied in various fields such as home appliances control, voice dialing, call routing, data entry in software, and speech-to-text processing. DSR has gained improvements in the recent days [35, 36]. A speech-supportive system [9] identified the articulatory errors of each dysarthric speaker, developed a speech recognition system that corrected the errors and developed a hidden Markov model-based speaker-adaptive speech synthesis system. Multiple qualitative and quantitative metrics [1] are considered to assess and evaluate dysarthric speech recognition systems. However, ASRs have not been widely used by people suffering from physical or neuro-motor disabilities such as dysarthria. In this chapter, I propose hybrid model based approach for dysarthric speech recognition.

6.2 Characteristics of Normal Speech and Dysarthric Speech

Normal speech is influenced by the state of the speaker such as happiness, anger, fear, or sadness. Speech characteristics are described by features such as loudness, intonation, and fluency. Loudness is associated with the articulation of utterances. Intonation is associated with the rise and fall of pitch. Speech is fluent when it is produced at normal rate without any interruption. Fluent speech is associated with various processes such as respiration, phonation, articulation, prosody and resonance. The waveform of a normal speech signal is shown in figure. When any of these processes gets affected, the speech thus becomes unrecognized. In the next section, I present the characteristics of dysarthric speech.

Dysarthria is a kind of neurological disorder that damages the control of motor speech articulation. When the speech mechanisms such as respiration, phonation, resonance, and articulation get affected, the speech becomes unintelligible. Some characteristics of dysarthric speech include breathy voice or non-voice, stuffy speech,

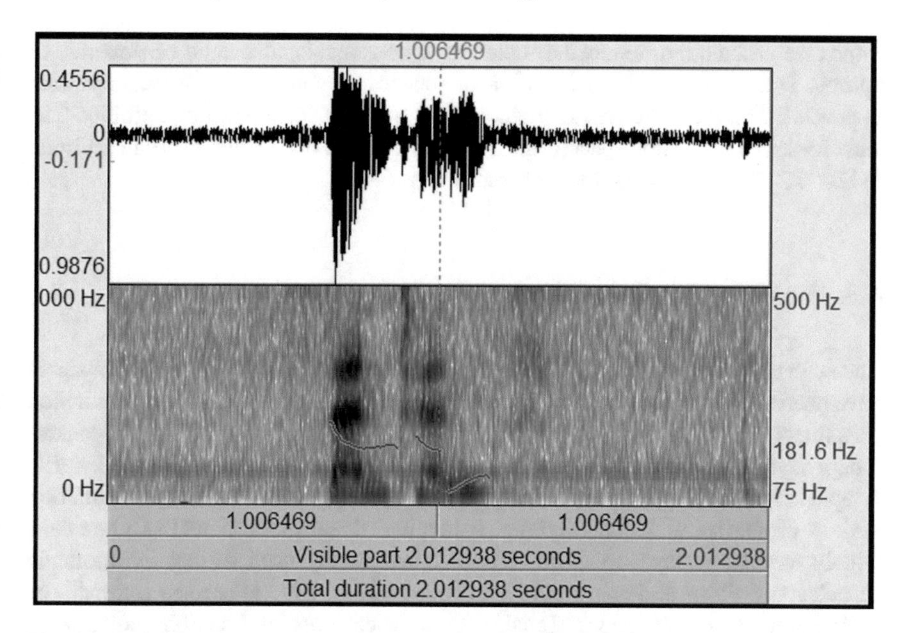

Fig. 6.1 Waveform of normal speech signal for the word 'Sentence'

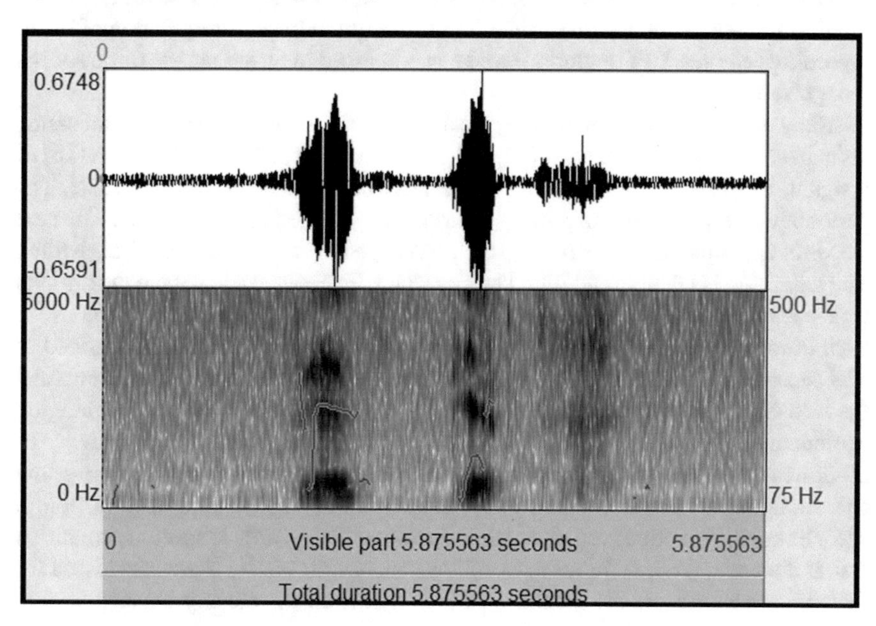

Fig. 6.2 Waveform of dysarthric speech signal for the word 'Sentence'

imprecise articulation, reduced or deletion of consonants, and rapid or slow rate of speech. Due to deficits in articulation of speech, no consistency in pronunciation is present. These characteristics reduce the dysarthric speaker's intelligibility. The waveform of a dysarthric speech signal is shown in Fig. 6.2. Distortions can be found in Fig. 6.2 and the same is not observed in Fig. 6.1.

6.3 Features for Dysarthric Speech Recognition

The main task in speech recognition is the feature extraction process. Several features have been used for dysarthric speech recognition task. Perceptual Linear Prediction uses linear predictive analysis for the approximation of the spectral shape. Speaker-independent information such as vocal tract characteristics can be represented by PLP. It represents the spectral shape better than linear prediction coding by approximating various properties of human hearing. Perceptual linear prediction (PLP) has been widely used as features in DSRs. Another extensively used feature in automatic speech recognition is linear predictive coding (LPC). LPC is used to estimate the basic parameters of a speech signal, such as vocal transfer function and formant frequencies. Linear Prediction Coding (LPC) is a linear predictor of order 'n' that predicts the sample at any given time 't' as a weighted linear interpolation of its 'n' preceding samples. LPC features have been employed in different dysarthric speech recognizers.

Many ASR systems which are typically used for disordered speech processing tasks have used MFCC features [12, 43]. MFCC is a cepstral representation [16] of a signal, in which the frequency of the bands is distributed across Mel scale. The periodogram estimate of the power spectrum is calculated for each frame. The next step is to apply the Mel filter bank to the power spectra, sum the energy in each filter. The logarithm is taken for all filter bank energies. The last step is to take the discrete cosine transform (DCT) for the log filter bank energies. The obtained 12 values for each frame are the Mel frequency cepstral coefficients. MFCC is thus described as Mel cepstrum with, the first 12 features are Mel-frequency cepstrum coefficients and the 13th feature is log energy. The remaining 26 features are the delta and acceleration coefficients. Thus, the speech signal of an utterance is represented as a sequence of 39-dimensional feature vectors with each feature vector representing a frame of the speech signal. Figure 6.3 shows the process of extracting feature vectors from a speech signal. The effectiveness of MFCC features for dysarthric speech recognition task is shown in [38]. A frame size of 25 ms and a shift of 10 ms are considered for feature extraction from the speech signal of an utterance in the experiments.

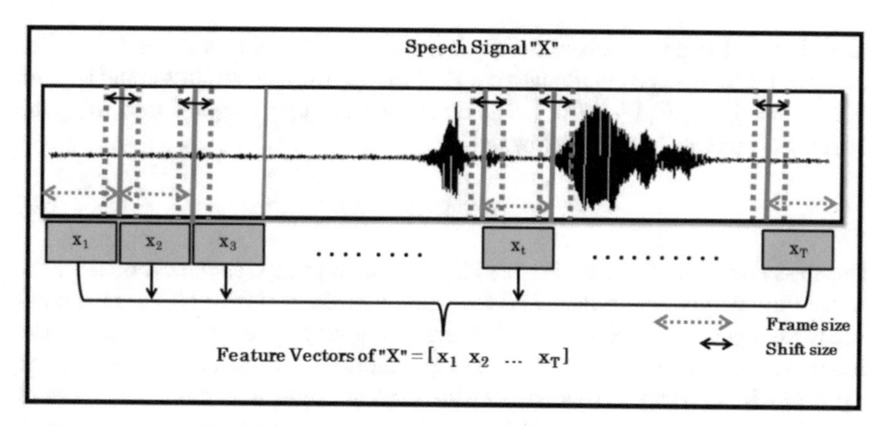

Fig. 6.3 Feature vectors extracted from a speech signal

6.4 Review of Model-Based Techniques for Dysarthric Speech Recognition

In this section, I present various model based approaches for recognizing dysarthric speech patterns. Model based approaches involve learning class boundaries to assign a class label to a speech signal [28, 29]. The two main model based approaches to learn a decision function from the examples are generative model based approaches and discriminative model based approaches. The generative model based approaches estimate the class conditional density for each example of each class using the examples of the respective class. The discriminative model based approaches construct the decision boundaries between the classes. I discuss generative model based approaches, discriminative model based approaches and hybrid model based approaches to recognize dysarthric speech in Sects. 6.4.1, 6.4.2, and 6.4.3 respectively.

6.4.1 Generative Model Based Approaches

Generative model based approaches uses the Bayes classifier to classify varying length patterns of speech. The Bayes classifier requires the likelihood of a speech utterance for each class. The Bayes decision rule is then applied to assign a class label to the example. Let C denote the total number of classes and $X = (x_1, x_2, \ldots, x_T)$ be a speech pattern. The likelihood of X for class c, P(X | c) is used to compute the corresponding posterior probability using the Bayes rule as follows:

$$P(c|X) = \frac{P(X|c)P(c)}{P(x)} \tag{6.1}$$

Here, P(c) is the prior probability of class c and P(X) is the evidence for X. The evidence is calculated by marginalizing $P(X|c)P(c)$ over all the classes and is given as $P(X) = \sum_{c'=1}^{C} P(X|c')P(c')$. The class label y is then assigned to X using the Bayes decision rule as given below in [2]

$$y = \text{argmax}_c P(c|X) \tag{6.2}$$

The class conditional density is generally computed using a suitable model for the probability distribution of data. Hidden Markov models (HMM) [30] are the commonly used generative models for classification of varying lengths patterns of short duration speech represented as sequences of feature vectors. In a speech recognition task, a HMM for a class is trained using the varying length observation sequences of multiple utterances belonging to that class. The HMM uses a finite number of states. The associated transitions jointly model the temporal and spectral variations in the speech signals. In HMM-based speech recognition, it is assumed that the sequence of observed feature vectors is generated by a sequence of hidden states with the first order Markovian assumption and using a left-to-right HMM. An HMM-based approach [26] was investigated using three types of feature vectors namely—the fast Fourier transform coefficients (FFT), linear predictive coefficients (LPC) and MFCCs.

An ASR system [14] studied the performance of hidden Markov model and support vector machine separately using the speech of one control speaker and three dysarthric speakers with low intelligibility level. Perceptual Linear Prediction (PLP) coefficients, energy, and MFCC coefficients were used as features. The performance was reported in terms of word recognition accuracy. The utterances that had missing consonants were not recognized by HMM-based recognition system. Stuttered utterances were recognized by SVM-based recognition system which assumed fixed word length. MFCC based ten-state ergodic model was found to be more accurate than the other two feature based model. Speech from three cerebral palsy speakers was used to test the model. The vocabulary consisted of 25 isolated words. Overall word recognition accuracy by the MFCC-based model was 92%. FFT based model and LPC based model produced 89% and 79.5% respectively. The database, called UA-Speech was constructed by Kim et al. [17] and it consists of speech data collected from dysarthric speakers with varied intelligibility levels. The database was used into build a speaker-adaptive and HMM-based speaker-dependent [39] using PLP coefficients. The performance of normal ASR system was used to evaluate the dysarthric speech of seven speakers. TIMIT database [45] was used to build the normal ASR. The respective mean percent word correct (PWC) scores for the speaker-adaptive dysarthric speech recognition and speaker-dependent ASR were 36.8% and 30.84% respectively.

A system [34] aimed to recognize and re-synthesize dysarthric speech and speech affected by Sound Substitution Disorder (SSD). Nemours database of American dysarthric speakers and disordered speech uttered by Acadian French speakers were used for investigation. The system used variable Hamming window size for each speaker. The size that produced the best recognition rate was used in the final system.

Techniques were introduced by the system to correct the voice of the speaker to enhance the intelligibility of dysarthric speech and SSD. MFCC features were used as an input to HMM-based recognition system. To measure the quality of speech, Perceptual Evaluation of the Speech Quality (PESQ) was used. The PESQ score measured was from 0.5 to 4.5. The larger the score, better the quality of speech. The authors proved the significance of the frame length of dysarthric speech. Besides, they demonstrated that the techniques used for synthesis improved the intelligibility of the speakers. The STARDUST (Speech Training and Recognition for Dysarthric Users of Assistive Technology) project [13] developed an HMM-based robust speech recognizer for three dysarthric speakers. The intelligibility level of the speakers was unknown. MFCC features were used for the experiments and the accuracy of the recognizer reported, was 94.33% for a ten-isolated word vocabulary. The study [22] presented in compared the accuracy of naive human listeners and the speaker-adapted HMM-based automatic speech recognition systems for isolated words of dysarthric TORGO database [32]. The speech obtained from six dysarthric speakers was used for evaluation. The mean word recognition accuracy was 68.39% for the speaker-adapted mono phone ASR systems using PLP. The mean percentage correct response of 14 naïve human listeners for isolated-word multiple-choice intelligibility test on the same speech was 79.78%. The performance of acoustic models based on MFCCs, Linear Predictive Coding-based Cepstral Coefficients (LPCCs), and PLP coefficients were compared. The mean word recognition accuracy of the baseline speaker-independent mono phone models based on PLP and MFCC coefficients with the zeroth order cepstral coefficients were 39.5% and 39.94% respectively. Word recognition accuracy of LPCC-based was 34.33%.

6.4.2 Discriminative Model Based Approaches

Artificial Neural Network (ANN) models are the discriminative non-linear models that do not make any assumption about the shape of distribution. An ANN model consists of interconnected processing units where each unit represents the model of an artificial neuron, and the interconnection between two units has a weight associated with it. One of the discriminative model based approach, used for many pattern classification applications is the artificial neural network (ANN). ANNs are more advantageous when compared to conventional HMM. They are effective in modeling non-linear data. In general, ANNs have many design parameters such as the number of hidden layers, the number of hidden neurons in each layer, and the activation function of the neuron. The output depends on these parameters.

An alternative approach with ANN-based model [15] used two multi-layer neural networks to recognize dysarthric speech of a male speaker with cerebral palsy for a ten-word vocabulary. One network used fast Fourier coefficients as input while the other network used formant frequencies. The results were compared with the intelligibility rating obtained by five human listeners. A recognition rate of 76.25% was obtained by ANN-based model produced a recognition rate of 76.25% and the recognition rate given by the five listeners for the dysarthric speaker's speech was 42.38%. An ANN-based speaker-independent (SI) ASR system [38] was developed to recognize dysarthric speech for a ten-digit vocabulary. Four MFCC feature sets such as 12 MFCCs (Set a), 12 MFCCs first derivatives (Set b), 12 MFCCs second derivatives (Set c) and Set d (Set a + Set b + Set c) were used. UA-Speech dysarthric corpus was used for experiments. The system using Set 'a' performed better than other sets of features. For the speaker independent DSR trained using the Mel cepstrum including 12 coefficients, the word error rate achieved was 68.38%. An assessment of dysarthric speech based on Elman back propagation network (EBN) [25] was studied for recognition of dysarthric speech. EBN is a fully connected network (recurrent network). The network used glottal features along with MFCC features as input. Evaluation was done on UA Speech database. The results of EBN proved effectiveness of the glottal features and voice excitation features. Speech recognition accuracy has been compared between the MFCC features and the glottal features using various modeling techniques. Combinations of articulatory knowledge [31] and modeling gave improved accuracies of recognition for individuals with speech impairments. These experiments included both theoretical and empirical representation of the vocal tract, with data obtained from MOCHA database [44] and from their own collection of dysarthric and normal speech, Dynamic Bayesian Network (DBN) showed improvement in phone recognition over conventional HMMs. However discriminative models such as Latent Dynamic Conditional Random Fields, SVM and ANNs showed greater improvements.

Many ANNs using MFCC features were employed by a dysarthric multi-networks speech recognizer (DM-NSR) model [37] to approximate the likelihood of the vocabulary. The experiments were carried out using UA-Speech database. When compared with dysarthric single-network speech recognizer (DS-NSR), the recognition rates of speaker-independent DM-NSR showed an improvement for low, moderate and high intelligibility dysarthric speech from 48.75%, 64.58%, 65.83% to 63.75%, 79.16%, and 83% respectively. Similarly, the mean recognition rate of speaker-dependent DM-NSR recorded an improvement from 64.94 to 80.83%. To effectively model the typical phonetic variation of dysarthric speech, the categorical distribution of Kullback Leibler divergence-hidden Markov model (KL-HMM) [19, 20] is trained on speech data from several dysarthric talkers using phoneme posterior probabilities obtained from a Deep Neural Network (DNN) acoustic model. Experiments were carried out in terms of word error rate (WER) on both 20 dysarthric and 10 control speakers. A comparison of results for DSR task using KL-HMM with that of GMM-HMM and DNN-HMM showed that KL-HMM provided significant improvement even for limited dysarthric training data was available.

6.4.3 Hybrid Model Based Approaches

An application of a 10-state ergodic HMM-ANN hybrid framework [27] with MFCC features was evaluated, to work as an assistive tool to recognize dysarthric speech spoken by three cerebral palsy subjects for a small vocabulary of isolated words. The recognizer produced better results than the conventional HMM-based classifier. Another hybrid system [33] uses the connectionist approach to estimate the severity level of dysarthria and speaker-dependent system to recognize dysarthric speech. With the use of new activation function based on class posterior distribution, a hierarchical structure of neural networks classifier estimates the severity level of dysarthria prior to dysarthric speech recognition. The input features consist of MFCCs and rhythm metrics based on duration characteristics of vocalic and intervocalic intervals and Pair wise Variability Index used with their raw and normalized measures. It was proven that the rhythm metrics influenced the observed durational differences among the dysarthric speakers. The Nemours database of American dysarthric speakers were used for the experiments. The results of hybrid system were compared with GMMs, single multi-layer perceptron (MLP) and standard hierarchical structure of MLPs.

A robust feature extraction method [24] was used instead of Mel frequency cepstral coefficients. The convolution bottle neck stacks various layers such as convolution layer, a sub sampling layer, a bottleneck layer to form a deep neural network. It has been shown that the feature extraction using CBN reduced the instability found in the dysarthric speech. The performance of the system was good comparing to conventional MFCC features. The model was evaluated on word recognition tasks for one person (male) with an articulation disorder. On evaluating with the deep belief network, over fitting and the lack of robustness found in dysarthric speech made the DBN to perform poor. Another hybrid model based approach was proposed to improve the recognition of dysarthric speech using the tempo adaptation of sonarants in dysarthric speech. The acoustic models such as GMM-HMM and DNN-HMM model based approaches were built using the speech data collected from control speakers. The authors compare speaker independent and speaker adapted recognition systems. The experimental results have shown that the speaker adapted recognition system has improved with tempo adaptation.

Polynomial sequence kernel based on dynamic time warping (DTW) [42] for recognition of isolated dysarthric speech utterances used support vector machine as the classifier. Speech data collected of 7 dysarthric speakers and 7 non-dysarthric speakers were used to build a speaker-dependent system. MFCC and their first order derivatives were used as features. Results were proven to be better than the standard DTW and HMM based approach. However, no significant improvement was seen, when the system was test with the unimpaired speech.

Another hybrid model based approach [4] used generative model driven (HMM based) features as fixed dimensional representation. This was given as input to the discriminative model based classifier such as support vector machine. Support vector machine was built in that likelihood score space. UA-Speech database was used for evaluation. The vocabulary used in the experiments included 19 computer commands and 10 digits. It outperformed the conventional HMM and deep neural network-hidden Markov model (DNN-HMM) based approaches.

6.4.4 Unsupervised Approaches

A post-classification posterior smoothing scheme [18] was proposed which refined the posterior of a test sample based on the posterior of the other test samples. Pitch stylization parameters, variance of pitch contour, normalized L0-norm, interquartile range of pitch and its delta and quantiles were used as features. On testing the performances of two data sets, namely TORGO database and NKI CCRT speech corpus [6], the authors have shown that the feature sets offered significant discrimination for binary intelligibility classification.

Three unsupervised acoustic model based approaches [41] were proposed for dysarthric speech recognition. They are: first, frame based Gaussian posterior-grams obtained from vector quantization, second, HMM of phone-like units called Acoustic Unit Descriptors (AUD) that are trained in an unsupervised manner, and, third, posterior-grams computed on the AUDs. 39-dimensional MFCC features were used. The authors utilized DOMOTICA-3 Flemish data set from ALADIN project [8] for evaluation. Seven out of nine speakers were considered to contribute dysarthric speech. Maximum F-score of 97.02% was achieved by Gaussian posterior-gram based representation.

6.5 Feature Learning Based Approach for Dysarthric Speech Recognition

A set of sequences of feature vectors extracted from the speech utterances belonging to that class is given as an input to the class-specific HMM. Each feature vector is formed using 39-dimensional Mel frequency cepstral coefficients. An HMM consists of states, and the observation sequence is given as a result of successive transitions from one state to another state. The states in an HMM correspond to the acoustic events. Information about the successive transitions between the states, which is the temporal modeling of the uttered word, is obtained by the model. The model also produces information about the stationary statistics of each state. This category of HMM is left to right, the transition to states with the smaller index is prohibited. The speech signal of an utterance with T number of frames is represented as $X = (x_1, x_2, \ldots, x_T)$ where x_t is a real valued observation at time t. Let N_c be the number of states in the model for a class c. Let π_i be the probability of initial state i. The condition $\sum_{i=1}^{N_c} \pi_i = 1$ should be satisfied for all initial states. Let a_{ij} denote transition probability from state i at time t to state j at time (t +1). Let $A = \{a_{ij}\}$ where $i, j = 1, 2, \ldots N_c$ denote transition probabilities among the various states. Let $s = \{s_1, s_2, \ldots, s_T\}$ denote the state sequence for observation sequence X. Let $b_j(x_t)$ denote the probability of x_t being generated by the model in state j at time t. The observation probability distribution is given as $B = \{b_j(x)\}$ where $i, j = 1, 2, \ldots N_c$. The continuous observation probability density for a state is modeled using the parameters such as weight, mean vector and covariance matrix of the qth component of

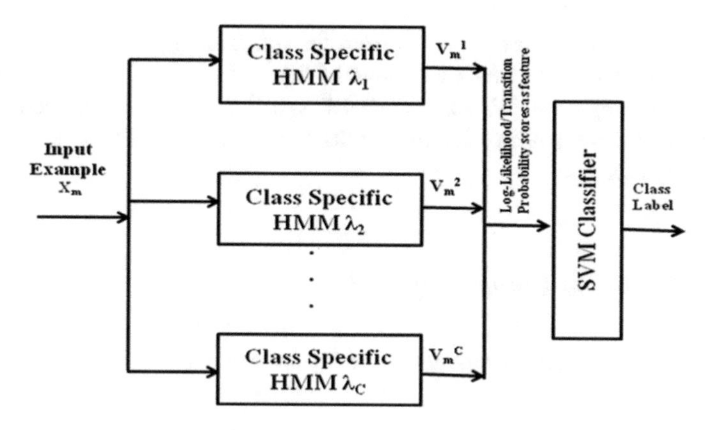

Fig. 6.4 Block diagram for the proposed approach

GMM of state j. The probability that an observation sequence of length T generated by the HMM $\{\lambda_c = A_c, B_c, \pi_c\}$ for a class c is obtained by summing the joint probability over all possible state sequences. A class label for Z is assigned based on the following decision logic:

$$y = \text{argmax}_c p(Z|\lambda_c)P(c) \tag{6.3}$$

where P(c) is the prior probability for class c. Baum-Welch method, an expectation maximization based method [21], or maximum-a posteriori method [11] is used for estimating the parameters of an HMM of each class. The log-likelihood scores and transition probability scores of class-specific HMMs are expected to show the nature of similarity. They measure how well a given example fits the class-specific models. For instance, the class-specific HMMs of examples from a class are expected to give high likelihood scores for an example that belongs to the same class and low likelihood scores for an example that does not belong to that class. HMM-based classifiers are robust only if sufficient training data is available. HMM-based approaches are not suitable for classifying the data of overlapping classes because a model is built for each class using the data belonging to only that class.

Perceptual characteristics of dysarthria include harsh vocal quality, imprecise consonant production, strained-strangled voice quality, hyper-nasality and slow rate. These characteristics of dysarthric speech lead to improper learning. I propose a Generative Model-Driven Feature Learning method for dysarthric speech recognition task in which the discriminative classifier is built in the generative model induced likelihood vector space and transition vector space. The generative model driven scores in the vector spaces provides the relevant information of a state or transition. When a discriminative classifier such as SVM is built in these spaces, the recognition accuracy is expected to be better. The block diagram is shown in Fig. 6.4.

Let $D = (X_1, X_2, \ldots, X_M)$ and $R = (R_1, R_2, \ldots, R_L)$ be the training and the reference data set of C classes respectively. Each pattern X_m corresponds to a sequence of feature vectors extracted from the mth training example. Let the sets R and D be completely distinct and none of the M examples in set D are chosen as reference examples. In this approach, class-specific HMM, λ_c is trained for each class C using the sequence of feature vectors of all examples of that class from the reference data set.

6.5.1 Likelihood Embedding

The log-likelihood scores for the sequence X_m are given by

$$V_m = \begin{pmatrix} \log p(X_m)|\lambda_1 \\ \log p(X_m)|\lambda_2 \\ . \\ . \\ \log p(X_m)|\lambda_C \end{pmatrix} \tag{6.4}$$

where λ_c is the class-specific HMM, modeled from the sequence of feature vectors of examples from the reference data set belonging to class c. The C class-specific HMMs, λ_c, c = 1, 2, .. C is used to obtain a C-dimensional log-likelihood vector representation for a training example or a test example. For a test pattern Z, the fixed dimensional vector representation is $z = [z_1, z_2, z_C]$ where $z_c = \log p(Z)|\lambda_c$. The log-likelihood scores have to be normalized from 0 to 1 as they are unbounded, before being presented to discriminative classifier.

6.5.2 Transition Probability Embedding

Let the model λ_c be trained on the sequences of feature vectors of class c from the reference data set, with the states $s^c = \{s_1^c, s_2^c, \ldots, s_{N_c}^c\}$, where N_c is the number of states in class c. Each sequence X_m is mapped onto a CK fixed dimensional vector where $K = \sum_{c=1}^{C} N_c^2$ and is defined as $V_m = [V_m^1, V_m^2, \ldots, V_m^C]$. Each V_m^c is obtained from the forward-backward algorithm, that represents the probability of passing from the state s_i^c at time t to s_j^c at time (t + 1). The fixed dimensional vector representation for a sequence Z is built in the similar manner. The values may or may not be normalized. In general there will be no necessity as they are the summation over the sequences of some fixed length. The fixed dimensional representations presented in Sects. 6.4.1 and 6.4.2 are used to build the discriminative classifier such as LE-SVM and TE-SVM.

6.5.3 Support Vector Machine

Discriminative model-based classifiers such as support vector machines focus on modeling the decision boundaries between classes. Support Vector Machine is a kernel-based maximum margin classifier, which has good generalization ability and it was proven to be effective in discriminating data between overlapping classes. Non-linear transformation is expected to lead to linear separability in the kernel feature space for data that are non-linearly separable in the input space. The LE-SVM and TE-SVM classifiers are modeled using the fixed dimensional representations V. An SVM [3, 7] is a binary classifier. Let the fixed dimensional vector representations of the training data V is $\{(V_i, y_i)\}_{i=1}^{M}$. Each V_i is the fixed dimensional vector representation and y_i is the respective class label. Let M_s denote the number of support vectors and $\{\alpha_i^*\}_{i=1}^{M_s}$ denote the optimal values of Lagrangian coefficients provided by the trained SVM. The discriminant function of the SVM for an input example Z is given by

$$D(Z) = \sum_{i=1}^{M_s} \alpha_i^* y_i K(Z, V_i) \tag{6.5}$$

The sign of D(Z) is used to determine the class of Z. SVM can be built using either one-against-the-one approach or one-against-the-rest approach for multi-class pattern classification. An input test pattern Z is classified by using the winner-takes-all strategy that applies the following decision rule:

$$Class\,label\,for\;Z = \text{argmax}_c D_c(Z) \tag{6.6}$$

6.6 Experimental Studies

6.6.1 Database for Recognizing Dysarthric Speech

The evaluation of the proposed isolated word dysarthric speech recognition is elaborated in this section. UA-Speech corpus [17] is the first open access database for dysarthric speech and the speech has been recorded from dysarthric speakers with varied intelligibility levels (a measure of how speech is understood), such as “very low (0–25%)”, “low (26–50%)”, “moderate (mid) (51–75%)” and “high (76–100%)”. The severity of dysarthria is classified based on the intelligibility level. If the speech is identified as “high intelligibility”, then the severity of dysarthria is low, and vice versa. The corpus comprises isolated words which include 10 computer commands, 10 digits, 26 radio alphabets, and 100 common words. The dysarthric speakers who contributed their speech data to this mostly suffer from spastic dysarthria. Speech intelligibility varies from 2 to 95%. Table 6.1 shows the intelligibility levels of the

Table 6.1 Intelligibility, severity levels of the speakers in the UA-speech corpus and the respective number of utterances used in this study

Speaker	Average Intelligibility (in %)	Intelligibility category	Dysarthric diagnosis	Number of utterances per class	Total number of utterances in reference set	Total number of utterance in training set	Total number of utterances in test set
M04	2	Very low	Spastic	5	145	116	29
F03	6	Very low	Spastic	7	203	153	50
M12	7	Very low	Mixed	6	174	135	39
M01	17	Very low	Spastic	4	116	82	34
M07	28	Low	Spastic	7	203	158	45
F02	29	Low	Spastic	7	203	154	49
M16	43	Low	Spastic	6	174	122	52
M05	58	Mid	Spastic	7	203	133	70
F04	62	Mid	Athetoid	7	203	159	44
M11	62	Mid	Athetoid	6	174	121	53
M09	86	High	Spastic	7	203	152	51
M14	90	High	Spastic	7	203	169	34
M10	93	High	Not sure	7	203	142	61
F05	95	High	Spastic	7	203	158	45
M08	95	High	Spastic	7	203	156	47

Table 6.2 Vocabulary-computer commands and digits

S. no	Word	S. no	Word	S. no	Word
1	Command	11	Paragraph	20	Zero
2	Backspace	12	Sentence	21	One
3	Delete	13	Paste	22	Two
4	Enter	14	Cut	23	Three
5	Tab	15	Copy	24	Four
6	Escape	16	Upward	25	Five
7	Alt	17	Downward	26	Six
8	Control	18	Left	27	Seven
9	Shift	19	Right	28	Eight
10	Line			29	Nine

speakers in the UA-Speech corpus and the number of utterances of each speaker used in this experiment.

The experimental studies in this paper are based on recognizing isolated word utterances of 29 classes from the UA-Speech vocabulary and are presented in Table 6.2. I have used the speech of 11 male and 4 female subjects suffering from dysarthria. A total of 2813 examples (i.e., 29 classes with 97 examples per class) were used such that 75% and 25% were used for training and testing respectively. The training and testing data set are disjoint and randomly chosen. The total number of utterances that was considered as reference set also contains 2813 examples. I have taken the sets R and D to be completely distinct and none of the M examples in set D were chosen as reference examples.

6.6.2 Results and Discussion

In this section, I present results of my studies on dysarthric speech recognition using conventional HMM and the proposed method. The left-to-right continuous density HMM-based system is considered in this study and the parameters are estimated using the maximum likelihood method. I consider diagonal covariance matrices for the state-specific Gaussian mixture models (GMMs). HMM for each class is built with the number of states N_c taking the value from the set $\{5, 6, 5, 4, 3, 5, 3, 5, 5, 5, 7, 5, 5, 3, 4, 5, 5, 5, 5, 4, 3, 3, 3, 4, 4, 4, 4, 5, 5\}$ for classes 1–29 respectively and Q, the number of components taking the value from the set $\{2, 3, 4, 5\}$ for the GMM of each state. I implemented HMM using Murphy's Matlab hidden Markov toolbox [23]. For the SVM classifier using the log-likelihood vectors (LE-SVM) and transition probability scores (TE-SVM), I considered the HMM with the same specifications as mentioned above. SVM classifier is modeled using LibSVM [5]. In this study, I consider the one-against-the-rest approach for 29-class speech recognition using

SVM classifier built in the log-likelihood vector spaces and transition vector spaces. The value of the trade-off parameter and the width of the Gaussian kernel used to build the SVM classifiers are empirically chosen. To assess the performance of the proposed model, Word Recognition Accuracy (WRA) is considered. It is given by

$$\text{WRA} = (\text{number of words correctly identified}/\text{number of words attempted}) \times 100 \tag{6.7}$$

Hidden Markov models have been used for modeling the temporal dynamics of varying length patterns. But for dysarthric speech recognition task, due to slurry, partial, slurry and weak articulation nature of speech, HMMs may not properly model them. The words are not recognized by conventional HMM, but recognized by the proposed methods such as LE-SVM and TE-SVM. As the log-likelihood and transition probability scores that are derived from HMMs provides the relevant information, the discriminative classifier such as SVM built in the HMM driven vector spaces give an improved performance when compared to the conventional HMM.

Figures 6.5, 6.6, 6.7 and 6.8 provide the word recognition accuracies of different speakers with intelligibility levels "very low", "low", "mid" and "high" respectively. Numbers within the brackets in the Figures correspond to the respective intelligibility levels. It is seen that the speech uttered by "very low" intelligible speakers are almost recognized. Another observation is that despite "high intelligibility", M10 could not produce good accuracy. The reason might be the lack of consistency present in the pronunciation. The recognition accuracies achieved for the conventional HMM, LE-SVM and TE-SVM for the dysarthric speech recognition task are given in Table 6.3. Due to the nature of speech and limited training data, conventional HMM did not perform well. But the discriminative classifiers such as LE-SVM and TE-SVM out-performed with overall recognition rate of 87.91 and 73.68%.

The average F-score achieved by LE-SVM and TE-SVM are 87.88% and 73.85% respectively. Though TE-SVM produced good results, it is important to handle the dimension which becomes huge.

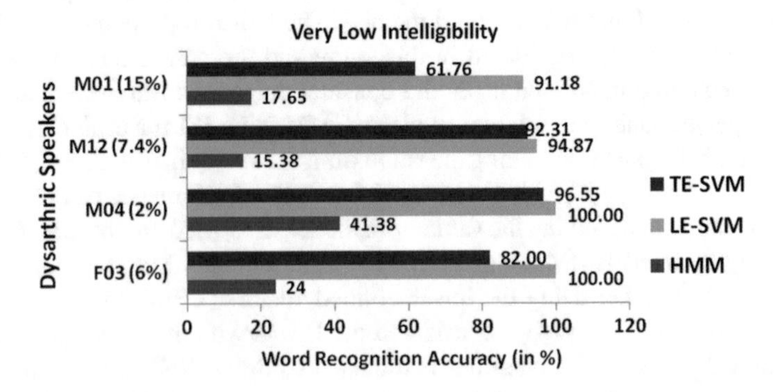

Fig. 6.5 Very low intelligibility speakers

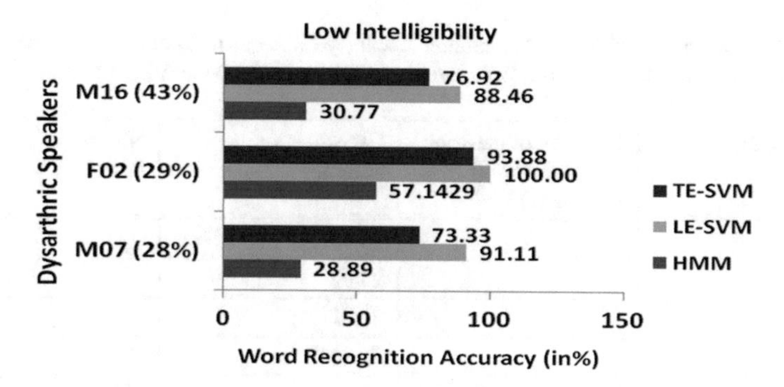

Fig. 6.6 Low intelligibility speakers

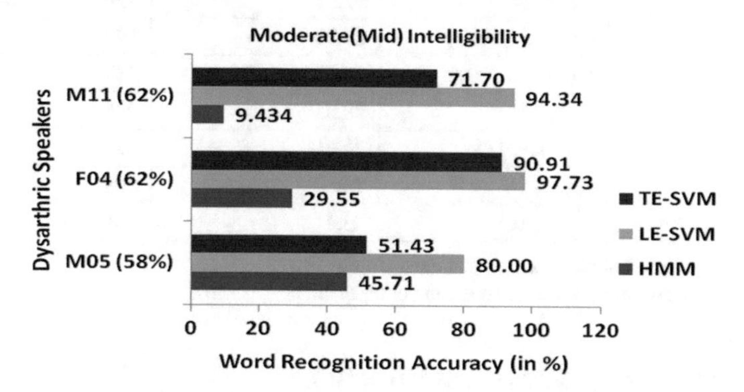

Fig. 6.7 Mid intelligibility speakers

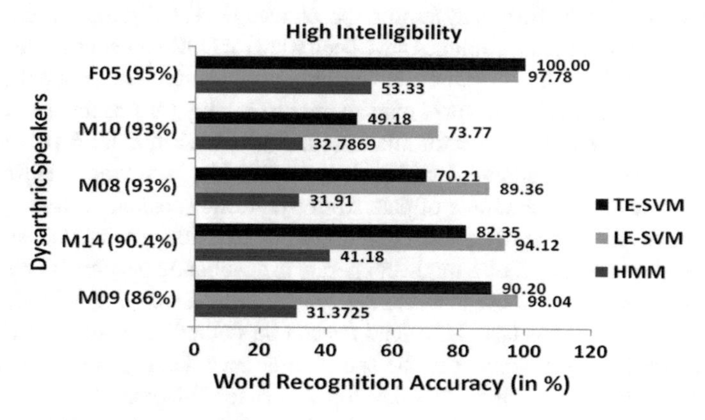

Fig. 6.8 High intelligibility speakers

Table 6.3 Comparison of Word Recognition Accuracy (WRA) (in %) obtained by conventional HMM and the generative model-driven classifiers such as LE-SVM and TE-SVM for dysarthric speech recognition task

Classification model	Q/(width of gaussian kernel, trade-off parameter)	Classification accuracy	Number of support vectors
HMM	2	28.3073	–
	3	28.02	–
	4	27.31	–
	5	26.8848	–
LE-SVM	(362.0387, 512)	87.62	165
	(2048, 90.5097)	**87.91**	156
	(2048, 90.5097)	87.34	143
	(2048, 90.5097)	87.48	137
TE-SVM	2, (64, 16)	70.55	240
	3, (2048, 16)	70.41	186
	4, (64, 16)	73.68	208
	5, (64, 16)	73.4	206

'Q' indicates number of components in state-specific GMMs
Bold denotes maximum performance among other results

6.7 Challenges and Future Directions

The performance of the current dysarthric speech recognizer still needs improvement. Many pattern recognition techniques have been tried for DSR task such HMM, ANN, and hybrid model based approaches. However, it is challenging to decide the classifier that suits this task. Future works may attempt to employ various feature selection methods to find the best feature for this task. It is also clear that DSR performance can be improved when enough training data is available. However, it is time consuming to collect more amounts of data from dysarthric speakers as they get tired soon. Besides, there are only a few open access dysarthric speech databases available. Several research institutes might cooperate in developing benchmark dysarthric speech databases. Most of the current research focuses on studying MFCC, LPCC, and PLP features for this task. New features can be studied to explore the relations to the dysarthric speech content of the speech utterance. Lastly, most of the DSRs are speaker-dependent systems. To develop a speaker-independent DSR, is quite challenging. So dysarthric speaker identification system can be implemented first, followed by a speaker-dependent dysarthric speech recognition system.

6.8 Conclusion

I have proposed generative model driven feature learning approaches with a discriminative classifier such as Likelihood Embedding-Support Vector Machine and Transition Embedding- Support Vector Machine. In LE-SVM and TE-SVM, a discriminative classifier is built in log likelihood vector space and a transition probability vector space respectively. The results obtained by each of the proposed approaches are better than the conventional HMM-based classification technique for all dysarthric speakers. LE-SVM captures a discriminative information which leads to give good recognition accuracy. Increase in the number of features in TE-SVM helps to give good performance but does not improve discrimination when compared to LE-SVM. The dimension increases in TE-SVM when there are more number of states and classes. Though there is loss of information in building models using partial or incomplete speech, the effective fixed dimensional representation formed using log likelihood probabilities given by HMMs, shows improved performance. This representation can also be applied for various sequential pattern classification tasks.

References

1. Asemi, A., Salim, S.S.B., Shahamiri, S.R., Asemi, A., Houshangi, N.: Adaptive neuro-fuzzy inference system for evaluating dysarthric automatic speech recognition (ASR) systems: a case study on MVML-based ASR. Soft Comput. 1–16 (2018)
2. Bishop, C.M.: Neural Networks for Pattern Recognition. Oxford University Press (1995)
3. Burges, C.J.: A tutorial on support vector machines for pattern recognition. Data Min. Knowl. Disc. **2**(2), 121–167 (1998)
4. Chandrakala, S., Rajeswari, N.: Representation learning based speech assistive system for persons with dysarthria. IEEE Trans. Neural Syst. Rehabil. Eng. **25**(9), 1510–1517 (2017)
5. Chang, C.-C., Lin, C.-J.: Libsvm: a library for support vector machines. ACM Trans. Intell. Syst. Technol. (TIST) **2**(3), 27 (2011)
6. Clapham, R.P., van der Molen, L., van Son, R., van den Brekel, M.W.M. , Hilgers, F.J.: NKI-CCRT corpus-speech intelligibility before and after advanced head and neck cancer treated with concomitant chemoradiotherapy. In: LREC, pp. 3350–3355. Citeseer (2012)
7. Cristianini, N., Shawe-Taylor, J.: An Introduction to Support Vector Machines and Other Kernel-Based Learning Methods. Cambridge University Press (2000)
8. De Pauw, G., Daelemans, W., Huyghe, J., Derboven, J., Vuegen, L., Van Den Broeck, B., Karsmakers, P., Vanrumste, B.: Self-taught Assistive Vocal Interfaces: An Overview of the ALADIN Project (2013)
9. Dhanalakshmi, M., Celin, T.M., Nagarajan, T., Vijayalakshmi, P.: Speech-input speech-output communication for dysarthric speakers using HMM-based speech recognition and adaptive synthesis system. Circ. Syst. Sig. Process. **37**(2), 674–703 (2018)
10. Duffy, J.R.: Motor speech disorders: clues to neurologic diagnosis. In: Parkinson's Disease and Movement Disorders, pp. 35–53. Springer (2000)
11. Gauvain, J.-L., Lee, C.-H.: Maximum a posteriori estimation for multivariate Gaussian mixture observations of Markov chains. IEEE Trans. Speech Audio Process. **2**(2), 291–298 (1994)
12. Godino-Llorente, J.I., Gomez-Vilda, P.: Automatic detection of voice impairments by means of short-term cepstral parameters and neural network based detectors. IEEE Trans. Biomed. Eng. **51**(2), 380–384 (2004)

13. Green, P.D., Carmichael, J., Hatzis, A., Enderby, P., Hawley, M.S., Parker, M.: Automatic speech recognition with sparse training data for dysarthric speakers. In: INTERSPEECH. Citeseer (2003)
14. Hasegawa-Johnson, M., Gunderson, J., Penman, A., Huang, T.: HMM-based and SVM-based recognition of the speech of talkers with spastic dysarthria. In: IEEE International Conference on Acoustics, Speech and Signal Processing, ICASSP, vol. 3, p. III. IEEE (2006)
15. Jayaram, G., Abdelhamied, K.: Experiments in dysarthric speech recognition using artificial neural networks. J. Rehabil. Res. Dev. **32**, 162 (1995)
16. Jurafsky, D., Martin, J.H.: Speech & Language Processing. Pearson Education India (2000)
17. Kim, H., Hasegawa-Johnson, M., Perlman, A., Gunderson, J., Huang, T.S., Watkin, K., Frame, S.: Dysarthric speech database for universal access research. In: INTERSPEECH, pp. 1741–1744 (2008)
18. Kim, J., Kumar, N., Tsiartas, A., Li, M., Narayanan, S.S.: Automatic intelligibility classification of sentence-level pathological speech. Comput. Speech Lang. **29**(1), 132–144 (2015)
19. Kim, M., Kim, Y., Yoo, J., Wang, J., Kim, H.: Regularized speaker adaptation of KL-HMM for dysarthric speech recognition. IEEE Trans. Neural Syst. Rehabil. Eng. **25**(9), 1581–1591 (2017)
20. Kim, M.J., Wang, J., Kim, H.: Dysarthric speech recognition using Kullback-Leibler divergence-based hidden Markov model. In: INTERSPEECH, pp. 2671–2675 (2016)
21. Lee, C., Rabiner, L., Pieraccini, R., Wilpon, J.: Acoustic modeling for large vocabulary speech recognition. Comput. Speech Lang. **4**(2), 127–165 (1990)
22. Mengistu, K.T., Rudzicz, F.: Comparing humans and automatic speech recognition systems in recognizing dysarthric speech. In: Advances in Artificial Intelligence, pp. 291–300 (2011)
23. Murphy, K.: Hidden Markov model HMM toolbox for Matlab. Online at http://www.ai.mit.edu/~murphyk/Software/HMM/hmm.html (1998)
24. Nakashika, T., Yoshioka, T., Takiguchi, T., Ariki, Y., Duffner, S., Garcia, C.: Dysarthric speech recognition using a convolutive bottleneck network. In: 12th International Conference on Signal Processing (ICSP), pp. 505–509. IEEE (2014)
25. Nidhyananthan, S.S., Shenbagalakshmi, V.O.: Assessment of dysarthric speech using Elman back propagation network (recurrent network) for speech recognition. Int. J. Speech Technol. **19**(3), 577–583 (2016)
26. Polur, P.D., Miller, G.E.: Experiments with fast Fourier transform, linear predictive and cepstral coefficients in dysarthric speech recognition algorithms using hidden Markov model. IEEE Trans. Neural Syst. Rehabil. Eng. **13**(4), 558–561 (2005)
27. Polur, P.D., Miller, G.E.: Investigation of an HMM/ANN hybrid structure in pattern recognition application using cepstral analysis of dysarthric (distorted) speech signals. Med. Eng. Phys. **28**(8), 741–748 (2006)
28. Povey, D., Burget, L., Agarwal, M., Akyazi, P., Kai, F., Ghoshal, A., Glembek, O., Goel, N., Karafiát, Martin, Rastrow, A., et al.: The subspace Gaussian mixture model—a structured model for speech recognition. Comput. Speech Lang. **25**(2), 404–439 (2011)
29. Rabiner, L.R., Juang, B.-H.: An introduction to hidden Markov models. IEEEASSP Mag. **3**(1), 4–16 (1986)
30. Rabiner, L.R., Juang, B.-H.: Fundamentals of Speech Recognition, vol. 14. PTR Prentice Hall Englewood Cliffs (1993)
31. Rudzicz, F.: Articulatory knowledge in the recognition of dysarthric speech. IEEE Trans. Audio Speech Lang. Process. **19**(4), 947–960 (2011)
32. Rudzicz, F., Namasivayam, A.K., Wolff, T.: The TORGO database of acoustic and articulatory speech from speakers with dysarthria. Lang. Resour. Eval. **46**(4), 523–541 (2012)
33. Selouani, S.-A., Dahmani, H., Amami, R., Hamam, H.: Using speech rhythm knowledge to improve dysarthric speech recognition. Int. J. Speech Technol. **15**(1), 57–64 (2012)
34. Selouani, S.-A., Yakoub, M.S., O'Shaughnessy, D.: Alternative speech communication system for persons with severe speech disorders. EURASIP J. Adv. Sig. Process. 6 (2009)

35. Seong, W.K., Kim, N.K., Ha, H.K., Kim, H.K.: A discriminative training method incorporating pronunciation variations for dysarthric automatic speech recognition. In: 2016 Asia-Pacific Signal and Information Processing Association Annual Summit and Conference (APSIPA), pp. 1–5. IEEE (2016)
36. Seong, W.K., Park, J.H., Kim, H.K.: Dysarthric Speech Recognition Error Correction Using Weighted Finite State Transducers Based on Context–Dependent Pronunciation Variation. Springer (2012)
37. Shahamiri, S.R., Salim, S.S.B.: A multi-views multi-learners approach towards dysarthric speech recognition using multi-nets artificial neural networks. IEEE Trans. Neural Syst. Rehabil. Eng. **22**(5), 1053–1063 (2014)
38. Shahamiri, S.R., Salim, S.S.B.: Artificial neural networks as speech recognisers for dysarthric speech: identifying the best-performing set of MFCC parameters and studying a speaker-independent approach. Adv. Eng. Inf. **28**(1), 102–110 (2014)
39. Sharma, H.V., Hasegawa-Johnson, M.: State-transition interpolation and map adaptation for HMM-based dysarthric speech recognition. In: Proceedings of the NAACL HLT 2010 Workshop on Speech and Language Processing for Assistive Technologies, pp. 72–79. Association for Computational Linguistics (2010)
40. Szatloczki, G., Hoffmann, I., Vincze, V., Kalman, J., Pakaski, M.: Speaking in Alzheimer's disease, is that an early sign? importance of changes in language abilities in Alzheimer's disease. Front. Aging Neurosci. **7**, 195 (2015)
41. Walter, O., Despotovic, V., Haeb-Umbach, R., Gemmeke, J., Ons, B.O.: An evaluation of unsupervised acoustic model training for a dysarthric speech interface. In: INTERSPEECH (2014)
42. Wan, V., Carmichael, J.: Polynomial dynamic time warping kernel support vector machines for dysarthric speech recognition with sparse training data. In: Ninth European Conference on Speech Communication and Technology (2005)
43. Wiśniewski, M., Kuniszyk-Jóźkowiak, W., Smołka, E., Suszyński, W.: Automatic detection of disorders in a continuous speech with the hidden Markov models approach. In: Computer Recognition Systems, vol. 2, pp. 445–453. Springer (2007)
44. Wrench, A.: The MOCHA-TIMIT articulatory database. Online at http://www.cstr.ed.ac.uk/research/projects/artic/mocha.html (1999)
45. Zue, V., Seneff, S., Glass, J.: Speech database development at MIT: TIMIT and beyond. Speech Commun. **9**(4), 351–356 (1990)

Chapter 7
Feasibility of Non-contact Smart Sensor-Based Falls Detection in a Residential Aged Care Environment

Ann Borda, Cathy Said, Cecily Gilbert, Frank Smolenaers, Michael McGrath and Kathleen Gray

Abstract Few studies of sensor-based falls detection devices have monitored older people in long-term care settings. The present investigation has addressed this gap by trialing the feasibility and acceptability of a non-contact smart sensor system (NCSSS) to monitor behaviour and detect falls in an Australian residential aged care facility (RAC). *Methods* This investigation was undertaken using a mixed methods approach, comprising three phases:

(1) Pilot study design and implementation at a RAC, using a purposive sampling approach;
(2) Study evaluation and post-pilot interviews; and
(3) Analysis and review of results.

Results Data was collected for four RAC participants over four weeks of the NCSSS pilot study. Numerous feasibility challenges were encountered, for example, in the installation configuration, placement of sensors for optimal detection, network and connectivity issues, and maintenance requirements. *Conclusion* The area of smart sensor technologies in falls monitoring and detection remains a relatively emergent field of investigation, and presently there are few real-life studies of NCSSS in an Australian RAC setting reported in the literature. This study confirmed that NCSSS technology may have a role in falls and behaviour monitoring of elderly residents in RAC and home environments. However, feasibility factors may affect implementation and adherence.

A. Borda (✉) · C. Gilbert · F. Smolenaers · K. Gray
Melbourne Medical School, Health and Biomedical Informatics Centre, University of Melbourne (FMDHS), Melbourne, Australia
e-mail: aborda@unimelb.edu.au

C. Said
Director of Physiotherapy Research, Austin Health/Physiotherapy, University of Melbourne (FMDHS), Melbourne, Australia

F. Smolenaers
Australian Centre for Health Innovation, Alfred Health, Melbourne, Australia

M. McGrath
Semantrix Pty Ltd., Melbourne, Australia

H. Costin et al. (eds.), *Recent Advances in Intelligent Assistive Technologies: Paradigms and Applications*, Intelligent Systems Reference Library 170,
https://doi.org/10.1007/978-3-030-30817-9_7

Keywords Falls detection · Falls monitoring · Smart sensors · Ambient assistive technology · Residential aged care · Patient safety

7.1 Introduction

7.1.1 Project Aims

The study aimed to investigate the feasibility of the implementation of a NCSSS to monitor behaviour and detect falls in a RAC. A hypothesis for the study is that automated smart monitoring using privacy preserving sensors can enable more accurate and prompt warning of falls and behaviour indicative of deterioration in older people living in residential care. Critically, the findings aimed to explore the acceptability of this system to older Australians living in RAC facilities.

7.1.2 Organisational Setting

A criterion for the study was that it should be positioned in a RAC in order to fill an identified gap among fall detection feasibility pilot studies in real-world settings [15]. This choice of location also simplified the recruitment of participants and, importantly, provided a managed environment in which the staff collaborated in the recruitment, in access, and in the ability of the team to gather feedback through interviews and discussions with staff and care-givers. The study collaboration comprised health informaticians, a specialist clinician, and the Technology partner, Semantrix Pty Ltd. (developer of the smart sensor technology).

7.1.3 Non-contact Smart Sensor System (NCSSS)

The study used a NCSSS implementation that was designed to run 24/7. Previous studies [14, 36] highlighted the documented challenges of body-worn sensors, for example, when trial participants detached the accelerometers intended to monitor their movements, or the awkwardness of placement on the body [34]. The prototype optical sensor technology was privacy-preserving, and had been specifically tailored for the study pilot by the technology partner, Semantrix Pty Ltd.

These smart sensors specifically utilise on-board cognitive processing and skeletal pose tracking to understand human movements in an indoor environment (see Fig. 7.1).

In this study, the NCSSS is calibrated to identify activities monitoring inferred skeletal movements and detecting, articulation movement patterns representing fall and other behaviours of interest or concern, such as patients attempting to exit a bed

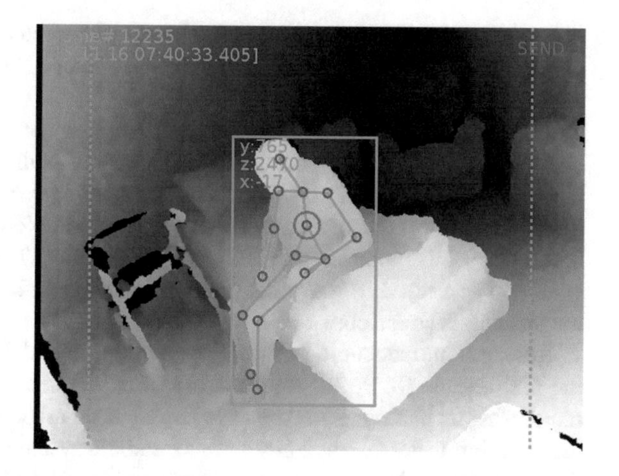

Fig. 7.1 Example of a depth image map generated by the smart sensor system. Image courtesy of Semantrix Pty Ltd.

unaided, or patients with low-blood pressure attempting to stand unaided after using a toilet. Unlike video surveillance, the cognitive processing of the system is carried out on-board the sensors and send "text summaries", such as "Patient X has fallen in the bathroom", which are transmitted as an alert to staff. For the purposes of the trial, the sensor prototypes were modified in two ways:

(a) Alerts were not transmitted to nursing staff due to the proof-of-concept status of the implementation and to reduce resource issues on the nursing staff in the event of test alerts;
(b) An additional depth map image was generated for further analysis post-trial to review correlations with other available information, such as pressure mat data, and to better understand characterisation of the system performance and validation of fall and behaviour detection algorithms.

There were 8 smart sensors ready for concurrent use, sufficient for four residential rooms with two sensors installed in each room. One sensor was positioned to monitor the bed area and the other sensor monitored the bathroom. Specifically, access to facilities like the shower and toilet and bed-exits were captured.

7.2 Background

7.2.1 Research Context

Falls and fall injury rates are markedly high in RAC facilities. The Australian Institute for Health and Welfare (AIHW) estimated that over 30% of community-dwelling older adults fall each year in Australia, with more than 83,000 hospitalisations for people aged 65+ [41]. This is nearly six times higher than the rate of falls in the home for this age group [5]. The rates are also exponentially high in terms of number of falls of patients when in hospital care. The AIWA reported about 34,000 falls resulted in patient harm in hospitals for 2015–16, at a rate of 3.2 falls per 1000 occupied bed days [3].

The burden of injury, disability, and loss of independence caused by falls will be an escalating cost to the health budget as the proportion of older people increases in Australia. Delays in rescue and treatment increases the pain and severity of a fall injury according to the Australian and New Zealand Falls Prevention Society.[1] A falls monitoring system can operate 24 × 7, automatically alerting staff when either a fall or a movement indicative of potential falls risk occurs, whether the resident is conscious or inactive.

Research into technologies for falls monitoring and detection has proliferated in the past decade [8, 9, 28]. In the array of experimental strategies to reduce falls, monitoring technologies to date have largely been associated with wearable devices (e.g. accelerometers), ambient sensors or camera systems [2, 20, 32, 33, 35]. Each of these approaches has disadvantages affecting efficacy and acceptability. The majority of these reports also focus on either simulated events or laboratory activities [25, 27]. Studies in real-life situations are less prevalent [19, 25]. The 2012 Cochrane review *Interventions for preventing falls in older people in care facilities and hospitals* [7] also did not identify relevant in situ trials of monitoring systems. It did, however, identify overall that multifactorial interventions in care facilities reduced the rate of falls of participants and risk, and recommended that there was a need for further trials testing sensor technology in place.

One notable field test of sensor monitoring is the University of Missouri's *TigerPlace* project, a purpose-built seniors' facility incorporating a mix of sensor types that continuously gather physiological data. These data enable detection of health status changes and trigger alerts to clinicians. A summary of two years' experience with 25 participants in the program noted benefits, including earlier detection of health deterioration and avoidance of crisis-related hospital visits [31]. Multimodal approaches to falls detection and prevention have also been piloted in other trials, e.g. the GAL@Home study [10] as a means to address more effectively the challenges of falls monitoring and to study the effectiveness of various sensor systems [10, 22].

Potter et al. [30] evaluated the performance of ambient depth sensors in capturing pre-fall activities and 16 actual falls in 13 hospital patients, half of whom were assessed as high risk using the hospital's fall screening tool. They concluded the sensor system provided realistic, usable data which could contribute to improved targeted interventions to prevent patient falls. In a subsequent trial, Potter et al. [29] compared falls and injury rates in two units in an acute hospital over time.

Both units were fitted with fall detection depth sensors; the intervention unit also had bed sensors to detect bed exits. The fall rate in the intervention unit was more than 50% lower than in the pre-sensor period, and also significantly lower than in the control unit with no bed-exit sensors. Another recent study on chair and bed exits using a battery-less and wireless wearable sensor system worn on the body

---

[1] Australian and New Zealand Falls Prevention Society. http://www.anzfallsprevention.org/info/. Accessed 1 May 2018.

was piloted using a scripted set of activities in a geriatric care unit of an Australian hospital showing preliminary results of high acceptance, but still requiring further trials on its efficacy [34].

Understanding feasibility and acceptability of these sensor interventions have been a characteristic of such studies (e.g. [10, 29, 30, 34, 38, 44]). Several recent reviews have also looked specifically at acceptability to older people of activity monitoring systems in the home [11, 16, 26]. Peek et al. [26] found that technology acceptance changes over time, however post-implementation research is still rare.

7.2.2 *Categorising Falls Studies*

The need for a common language describing technologies used for falls detection, prevention, prediction, and assessment has been recognised by falls researchers over the past 20 years [4, 18]. In response, The European Commission-funded FARSEEING consortium devised the *FARSEEING Taxonomy of Technologies* as a tool to enable falls studies and interventions to be characterised and classified using a standard 'language'.

The taxonomy employs five broad domains: *Approach, Base, Components, Descriptors,* and *Evaluation*. Within these, subdomains and categories have been defined.

Clarification, testing and refinement occurred through a consensus process among the partners.[2]

Table 7.1 provides a list of the FARSEEING domains and sub-domains which are relevant to the current study.

The Taxonomy Development Group also notes that this product built on the lessons of the STARE-HI project [6] and is compatible with its categories for a reporting study context. The FARSEEING Consortium has further developed a database of real-world fall events, with the objective of gathering actual falls case reports recorded by sensors.

7.3 Methods

7.3.1 *Study Design*

The research project utilised a mixed methods approach comprising three phases, namely:

---

[2]The full version of the taxonomy and the accompanying handbook are available on the FARSEEING project website: farseeingresearch.eu. The online application can be viewed at http://taxonomy.farseeingresearch.eu.

Table 7.1 Study mapped to FARSEEING taxonomy

Domain A: Approach	Domain D: Descriptors
(A1) Primary aim of the intervention reported in the study • (A1.3) To detect falls • (A1.6) To monitor function/physical activity and participation in activity • (A1.9) To undertake technological development **(A2) Study design Type of study being conducted** • (A2.1.6) Feasibility study • (A2.1.4) Quasi-experiments • (A2.1.7) Mixed methods **(A3) Main Selection Criteria** • (A3.2.1) Age Group • (A3.2.5) Previous fall history **(A3.3) Chronic diseases, symptoms, impairments** • (A3.3.8) Gait and/or balance impairment • (A3.3.5) Dementia, cognitive impairment **(A3.6) Specific groups excluded** • (A3.6.3) Other specified exclusion	**(D1) Technology location** • (D1.2) Located in the environment **(D2) Technology type** • (D2.3) SE (Sensor) • (D2.3.11) Image Processor **(D3) Functionality** • (D3.1) Alert • (D3.2) Monitoring • (D3.7.1) Automatic **(D4) Method** • (D4.3) Visual • (D4.4) Other **(D5) Initial training/Instruction** • (D5.1) Duration of contacts • (D5.2) Intensity of intervention **(D6) Supervision/Follow-up contacts** • (D6.1) Frequency of contacts **(D7) Intervention utilisation:** • (D7.1) Frequency of use • (D7.2) Timescale of use
Domain B: Base	
(B2) Residential long-term (nursing) care facilities (non-acute) • (B2.1) Long-term nursing care facilities (HP.2.1) **(B9) Educational research settings not elsewhere classified** (e.g. Research Laboratory)	
Domain C: Components of outcome measures	Domain E: Evaluation
(C1) Outcome measures recorded/carried out by • (C1.1) Professionals • (C1.3) Researcher • (C1.5) Automated **(C2) Medium of outcome measurement** • (C2.1) Face-to-face • (C2.2) Paper-based • (C2.4) Technology led **(C3) Outcome measurement method** • (C3.1) Observation • (C3.2) Clinical assessment (e.g. pre-trial) • (C3.4) Non-validated assessment tool **(C4) Outcome measurement implementation** • (C4.3) Attitudes towards technology • (C4.16) Other outcome measures	**(E4) Ethics** • (E4.1) Ethical issues **(E5) Stakeholder technology perceptions** • (E5.1) Physical • (E5.2) Usability • (E5.3) Function • (E5.4) (Technical) support • (E5.5) Data security and protection **(E6) Participant perceptions** • (E6.1.2) Usability • (E6.1.3) Privacy • (E6.1.4) Function • (E6.1.5) Human interaction • (E6.1.8) Sustainability • (E6.1.10) Control **(E6.2) Service satisfaction** • (E6.2.1) Service satisfaction **(E7) Participant adherence** • (E7.1) Adherence

- Pilot study design and implementation at a RAC, using a purposive sampling approach;
- Pilot evaluation and post-pilot interviews; and
- Analysis and review of results.

Using a mixed methods approach for the analysis is also appropriate to this sample size and for correlating data outputs across qualitative and quantitative data collection instruments. Given the lack of real-world falls monitoring data available for Australian aged care facilities, it was deemed valuable to test the recently-developed smart privacy-preserving sensor technology in this pilot study.

The study pilot took place in a 170-place aged care facility located in a southern suburb of Melbourne. The site had 200 staff supporting 165 beds; 144 beds in single rooms and 10 in double rooms. A day-centre also provided support for families and carers. The pilot was initially designed for an eight-week duration period at the RAC.

7.3.2 *Participants*

Following ethics approval through the University of Melbourne, the study aimed to recruit four participants. Participants were recruited from a specific unit within the RAC. Baseline inclusion criteria for participants for the study were as follows:

- Aged 65+
- Previous falls history (e.g. two or more in past 6 months)
- Able to walk either independently or with staff assistance (+/− gait aid).

Exclusion criteria for participants focused primarily on immobility and extent of functional decline. Criteria were as follows:

- Bed bound, or requiring hoist for all transfers
- Residents currently receiving palliative care
- Residents who have had no falls in previous 12 months.

A purposive sampling technique was used in the study. RAC staff conducted a falls assessment during the selection stage. The staff agreed with a nine-point checklist in order to create a profile of potential candidates, and based on the inclusion criteria as outlined below in 4.1: Table 7.2. The checklist covered such assessment areas as age of patient, falls risk, medical diagnosis, mobility, and level of care requirements.

Data on potential candidates were drawn from a computerised clinical care package (SARAH), used with a suite of assessments, which the RAC utilised for the selection process. The Residential Care Manager initially screened the facility's list of occupants to identify eligible residents. The initial tranche of potential participants internally selected by the RAC was not provided to the research team. The names of four participant residents were put forward in a final selection list based on the assessments, with a leaning towards residents profiled in a high falls risk category and with a previous falls history (Table 7.2). Two residents lived in the general section of

Table 7.2 Summary of participant characteristics

Assessment	Resident 1	Resident 2	Resident 3	Resident 4
Medical diagnosis	Parkinson disease	Diabetes	Vascular dementia	Type 2 diabetes
	Dementia	Vertigo	Osteoarthritis	Pacemaker
	Type 2 diabetes	Atrial fibrillation	Lumbar vertebral crush # L2–L3	Atrial fibrillation
	Meniere's disease	Pacemaker	Chronic obstructive airways disease (COAD)	Oedema
		Lumbar laminectomy	Chronic obstructive pulmonary disease (COPD)	Short term memory loss (STML)
Falls risk assessment score	HIGH falls risk	HIGH falls risk	HIGH falls risk	HIGH falls risk
Previous falls history: in summary (not detailed)	Not available	5 falls in the last 6 months; getting out of bed, misjudging the furniture, attempting to stand up, trying to get dressed	13 falls in the last 6 months; attempting ADLs without assistance, attempting to go to the toilet, fall out of bed, found on floor several times, fall with a frame	1 fall in the last 6 months
Cognition score: Psychogeriatric assessment scale (PAS)	PAS = 11	PAS = 7	PAS = 15	PAS = 3
Mobility: whether he or she is independently mobile, or needs assistance from staff. If yes, to what extent?	Assist ×1	Assist ×1	Transfers assist ×2 and 4 wheel walkers in room	Supervision ×1

(continued)

Table 7.2 (continued)

Assessment	Resident 1	Resident 2	Resident 3	Resident 4
Does he or she use a gait aid?	4 wheel walker	4 wheel walker	Wheelchair beyond room	4 wheel walker
Activities of Daily Living (ADL) score?	High	Low	High	Medium
Is he or she assessed as high care, low care, or other?	High	High	High	High

the RAC facility and two were from the dementia care unit. Domain experts advising the research team suggest that these numbers, though small, would be adequate for this exploratory study to gather useful data and responses. Similar cohort sizes can be found in studies by Gietzelt et al. [14] and Suzuki et al. [37], for example.

People with dementia were specifically targeted for inclusion as they are acknowledged to have a higher risk of falls and fall-related injuries [1, 12, 24, 44]. The advanced age range of the participants is further associated with higher incident rates of falls [24, 40, 43].

The RAC Manager coordinated the approach to potential candidates or their responsible person (for people with cognitive impairment) and sought their written consent to participate. In addition to written and verbal information about the project, potential participants were provided with a short demonstrator session, conducted by the Technology partner, which preceded the start of the pilot.

Nursing staff agreement fell under the broader consent of the project undertaking by the RAC management. A research team member and the Technology partner also provided a broad summary of the study's interventions in a group session with staff at the RAC. Staff participants were specifically asked to: (a) carry out their usual activities with a participating resident during the observation period and (b) to respond to room alerts (e.g. pressure mat) concerning participating residents with the usual urgency.

7.3.3 Study Flow

7.3.3.1 Sensor System Lab Test

The study was preceded by laboratory testing of the smart sensor system. This involved a healthy volunteer simulating fall behaviours associated, for instance, with standing, sitting, and getting into and out of bed. The precision of the system in tracking a person, for instance, the effectiveness in capturing joint points and articulation movements indicating predictive fall behaviours were particularly targeted in the laboratory simulation. The latter further informed the positioning of sensors (e.g. optimal distance) in a monitoring environment. A sensor database was concurrently developed to handle sensor data for the pilot.

7.3.3.2 Pilot Study Period

A test of the sensor system took place within the bedroom of one of the participants when the resident was at a meal. This allowed the Technology partner to check for optimal placement of the individual sensors in the bedroom and bathroom as informed by the lab test. This pre-trial also provided an opportunity to test the sensors in situ with the sensor database, so that data flows, data capture, connectivity, and other issues might be identified at this stage.

The roll-out of the first pilot trial took place after the resolution of a number of issues arising at the testing stage (see Sect. 4.2: Unexpected events). The pilot roll-out was staggered so that each participant's room was fitted out and tested before the next roll-out to the subsequent participant's room. This was to allow for any customisations or adjustments that needed to take place, considering each room had slightly different configurations.

The pilot study period was planned for eight weeks in duration including the set-up of the sensor system, the smart sensor monitoring of patients, and the dismantling of the system.

7.3.3.3 Evaluation

Evaluation of the system was intended to take place pre-and post pilot study, with the use of a brief interview of participating residents and RAC staff who were working in the unit where the sensors were installed. The pre-study interview aimed to gather participants' views about the benefits and drawbacks of the monitoring and falls detection technology, for example.

A brief post-study interview was conducted to discuss the experience of staff during the study and solicit views about monitoring and falls detection technology. The interviews were designed to take no more than 30 min, and were to be audio-recorded to ensure accurate capture of the feedback data.

7.3.4 Outcome Measures

Outcome measures fell into two primary areas; system efficacy and technology acceptance.

Efficacy was described as the ability to capture data for, and alert staff to, a falls event or to behaviour potentially leading to a fall. Acceptance of the technology was interpreted as the level of adherence to the monitoring and willingness to participate by study residents (or carers), and by RAC staff.

7.3.5 Data Acquisition and Measurement

The smart sensor system was intended to alert for falls, trips, or slips in real-time. A dataflow system was architected to inform necessary infrastructure requirements for sensor data generation and capture, secure storage, and retrieval for analysis.

The 24/7 'movement' monitoring of resident rooms was enabled with approximately 18–25 GB of data generated per day/per room. Resulting files averaged 21 GB of sensor data blocks comprised of high compression files (.tar.gz), saved onto a secure, dedicated server.

According to the Residential Manager, there are 17,000 alerts per month (many of which are not associated with falls) of which staff respond to 75% of calls on average within 5 min. For vulnerable residents, pressure mats are situated on the floor and also in the bed; when the pressure changes (i.e. a patient moves off the bed or onto the floor) this triggers a Nurse Call alarm.

Data was provided by the RAC for call stations and pressure mats for the study period. This period covered September, October, and November 2015 inclusive. It was intended post-trial that correlations across these data and smart sensor data (e.g. visualisation of skeletal movement patterns) could be made to validate fall events and/or fall behaviour patterns, and to refine detection algorithms.

7.3.6 Data Analysis

Data analysis was planned across several areas of data collection as outlined in 3.5, namely:

(a) An analysis of the sensor data collected from the 8 sensors installed in the bedrooms and bathrooms for the four participants. In each case, the movement data was collected as de-identified data to be analysed off-site from the RAC. Collected data was to be analysed to verify if equivalent patient events had been detected from other data systems, namely pressure-mat system data and nurse call alert data. This analysis was to explore not only individual fall detection results, but to aggregate results for comparative purposes and to understand anomalies and discernible patterns.

(b) A second analysis was planned to determine if the sensor data correlated with the pressure-mat system data and nurse call alert data that were available for each participant room during the time of the pilot. These correlations would allow the investigators to understand, in the eventuality of a fall, if the sensor system captured the fall event at the same time as other systems in place. The correlations could also be analysed in terms of predictive fall behaviours. For example, a nurse call alert may be timed with the occurrence of a fall incident, in which case the smart sensor depth images locked at that point in time could be examined, to understand the efficacy of the system in capturing and alerting predictive fall behaviour.

(c) Thematic content analysis was utilised for the interview instruments. This would enable responses by participants, carers and facility staff to be examined in order to identify where views converged or differed.

7.4 Results

7.4.1 Demographic and Coverage Data

Four male residents with an average age of 87 years were identified by RAC staff as potential participants in the pilot study. Each of the patients resided in a single room with an en suite bathroom in the RAC facility. In each case, the participants were assessed as high care: three of the patients used a gait aid in the form of a four-wheeled walking frame, and one participant used a wheelchair when moving beyond the confines of his room. Three of the patients required staff assistance for mobility, and one was supervised. Health conditions varied across the participants, but each had various co-morbidities. Three of the four participants were diagnosed with a diabetic condition and breathing difficulties. Two participants were diagnosed with clinical dementia. Each patient received an assessment using a Psychogeriatric Assessment Scale (PAS) in which a higher score suggested cognitive decline in three of the participants. At least two of the participants were identified as having a high falls risk - one patient was known to have fallen 13 times within six months, largely trying to undertake Activities of Daily Living (ADL) routines without assistance. Another participant had a recorded five falls within six months. Patient assessments are summarised in Table 7.2.

Each of the participant rooms was fitted with a smart sensor in the bedroom in a location in which the system could lock onto the resident and the bed, and a second sensor in the bathroom in which the system could lock onto a resident and the toilet area.

Three of the four residents took part for the full monitoring period of the pilot. Technical difficulties prevented a fourth participant resident to be observed by the system. No participants or individual carers withdrew from the study or identified privacy concerns over the course of the pilot.

7.4.2 *Unexpected Events*

7.4.2.1 Lab Testing

During the lab-testing phase of the smart sensor system, a healthy volunteer was used to test the functionality of the system. The volunteer was not representative of the target group, so issues associated with tracking of an older person, such as an older person leaning towards a gait aid, were not identified until real testing with residents of the RAC. Lab testing informed optimal range of the sensor system, and the RAC initially arranged for the Technology partner to use an empty suite in the residents' area to live-test optimal placement of the sensors in bedroom and bathroom areas.

The limitation in range and quality of the commercial pose-tracking component of the sensor system turned out to be more significant than anticipated in the set-up trial. This limitation was challenged when the sensors were moved outside of the planned range following the identification of network interferences.

7.4.2.2 Network Connectivity

The Technology partner found that the RAC network environment was configured differently to what was described in technical meetings prior to installation. For instance at the time of installation, it was discovered that the participant rooms had data points but these had not been cabled to the core network.

Hence there were significant issues at the time of initial beta-testing and installation, in terms of connectivity in the participant rooms. The Technology partner was unable to establish, maintain, or reconnect sensors without a workaround for the wireless network issues, with a temporary network solution, such attempts to use Power over Ethernet (PoE) data over powerline and later local wireless repeater nodes mounted in each room. The final solution used sensors connected wirelessly to the repeaters which in turn connected via the Ethernet based phone data cabling instead of the pure data cabling to the RAC network. While this path was much less direct than cabling directly to the RAC network, it enabled connection and data capture from the sensors to the data centre.

7.4.2.3 Interference

Another unexpected issue was due to TV radio-frequency interference. Sensors were installed into actual patient rooms (e.g. when a patient was out for a meal), and subsequently were moved as problems with radio frequency interference were found. This introduced issues with tracking due to the need to increase the distance of the sensors from the bed and bathroom areas being monitored.

Moving the sensors out of the optimal bed sensing range of 1.5 metres from the TV allowed 80% reliable communications, but consequently moving the sensors outside their notionally stable 4 metres detection range impacted the skeletal tracking performance and captured high levels of 'noise'. Initially the detection sensitivity threshold of the sensors was set to 'low' in which the movement range and movement velocity, among other variables, are captured with less filtering so more data is captured for predictive analysis. However given the issues, it was necessary to increase the threshold to reduce the amount of data being continuously generated. As a result of the preliminary review of events during installation testing, the Technology partner re-calibrated the smart sensors to enable "high" and "medium" filtering levels of falls detection in order to reduce excess information capture and to reduce the amount of data generation. Moving to medium (e.g. near fall event or partially unrestrained fall) and high (e.g. high impact or unrestrained fall) event detection settings meant only more obvious or unrestrained movements with fewer subtle variations were captured and less data overall was generated.

Lastly in some rooms, it was discovered that the sensors' power-supplies were being turned off. For instance, one sensor went offline when the power supply suddenly turned off. An elderly resident on a floor bed was found to have disconnected cables whilst curiously investigating the sensor system.

7.4.2.4 Organisational Change

During the study period, there were changes in staff, including the RAC Residential Care Manager who provided support in the original study grant proposal. This and other organisational changes during the study resulted in re-prioritisations that impacted on aspects of the pilot implementation, for example, such as networking and connectivity infrastructure to be made available in some of the participant rooms.

7.4.3 Study Findings and Outcome Data

7.4.3.1 Summary Findings

One known patient fall incident occurred during the trial period. This was not recorded as the system was still in the process of being commissioned. That particular patient room was identified with a faulty network cable connection which prevented its sensor connecting during that time.

Maximum observation period of an individual participant was intended to be eight weeks (duration of the pilot) but this was reduced to a four-week monitoring trial period due to challenges with the network and technical issues during set up. In total, 8 sensors were installed in 4 residents' units, and generated data for a combined total of 122 days. Three participants were monitored for periods varying between 5 and 22 days during the 4-week monitoring period (the average period was 17.5 days). Technical difficulties in implementing sensor connectivity prevented a fourth participant resident to be observed by the system.

Overall, there was only one instance of 2 consecutive days achieving the desired outcome of continuous uninterrupted monitoring. Interruptions to monitoring included network interference and power supply interruptions as documented previously. During the consecutive two-day monitoring period, no falls or near fall events were detected by the sensors. Post-analysis of the results within this sample time-period aligns with the monitoring data by the institution's nurse-call system and individual pressure mat data.

7.4.3.2 Participant Findings and Outcomes

Resident participants were specifically asked to give consent that, if a fall, trip or slip occurred, a text summary of the event and a depth map image may be collected, de-identified and stored securely for analysis. As participants were largely unaware of sensors due to cognitive decline, there was formal agreement by carers that this information could be collected.

Participants were similarly approached to note any fall, trip or slip that occurs during the study period in a 'falls calendar'. This similarly did not occur in the trial period as participants were not functionally able to document falls themselves, or to participate in pre- and post-interviews. Subsequently, there was consent to approach nursing staff to provide alert call data and any recorded data which might indicate an actual fall or near fall event. Post-study interviews took place with staff working in the areas where the sensors were installed from the start to the end of the trial, to discuss their perceptions and experience.

7.4.3.3 Stakeholder Perceptions

While participants and carers were informed that they were unlikely to benefit directly from this study, the family of one resident expressed their hope to the Technology partner team member that "the research could eventually lead to a system that did improve early detection."

Notwithstanding, staff felt that they were contributing to research that may help older people avoid falls at home, in hospital, or in residential care facilities, such as in the present study. Perceptions of key staff involved in the trial were documented in a post-trial interview by phone. A selection of questions and de-identified responses are provided in Table 7.3.

7.4.4 Unexpected Observations

Despite continuous problem identification with the network and work-arounds in the sensor system deployment, the staff retained a positive approach towards the project and willingness to participate. This, perhaps, emphasises the importance of

Table 7.3 Post-trial interview questions and de-identified responses

Interview question	*Response*
Did the installation of the sensors alter the staff's relationship or frequency of contact with the residents, do you think?	*No. The ones that obviously had the sensors were already needing a bit more assistance and regular checking than some other residents because of their high falls risk anyway*
Overall your feelings about using a technology like this for monitoring behavior and falls detection—what would you say?	*I'd strongly agree… That it's worth persisting with. 'Cos the thing is, obviously when it's up and working correctly, and not falling out of the system, it will give us a good indication of the events leading up to a fall. Yes, it would be very good and useful*
Do you think the installation, having the sensors in the rooms, made any difference at all to the way staff and residents interacted?	*I really don't think it had any negative impacts on anybody*
What's your feelings then about the value of this technology in monitoring movement and behavior and falls in a nursing home environment?	*… So if there's anything, any research or any information out there that can assist us to be a bit more proactive, we're all for it because we want to keep our people safe*
Did you have any hesitation about the visual side, so far as the residents were concerned, that they might be daunted?	*No, they did tell us it wasn't actually…like, it's not a visual of their face, it's just a—like, stick figures basically, kind of thing. So that's good for privacy and confidentiality, obviously*
When you think about things like the pressure mat system and the nurse call buttons, do you have any mental comparison of the effectiveness of each of those sorts of systems in comparison with the sensors?	*Yes, but there's a difficulty about the timeliness of the staff. Yes, we're aware that the bell goes off, yes we're aware that people are on the move, but it's what's also happening in the building at that particular moment in time*
Can you see it having wider applicability in a place like [the RAC] or do you think it's more useful where people aren't able to readily communicate themselves?	*I think it would be useful in [the RAC], overall, especially for people that obviously have dementia and can't articulate how the fall happened. It is a good measure of knowing what was leading to the fall and how they did it…*
Has being present and part of the sensor trial at [the RAC] changed your view in any way about whether sensors are the way to go when it comes to falls detection and falls prediction?	*It's exciting where the research is taking us… I just think that that's taking it to the next level, and it's really really exciting, to ultimately keep our people safe*

falls risk to RACs—i.e. aged care professionals are willing to look at innovative solutions. Relatives of the residents, nursing and maintenance staff were also interested, supportive, and helpful throughout the trial. This is borne out by other falls studies which suggest that although there is a limited pool of knowledge about clinicians' attitudes toward health technology use in falls avoidance, the attitudes are generally supportive where they have been documented [17, 39].

Self-reporting by participating residents were also part of the original qualitative study design so that interviews were focused on staff and carers in the context of dementia participants not being fully aware of the pilot [1]. In working with dementia patients in the pilot, the study follows the hypothesis put forward by Stucki et al. [36], Nijhof et al. [23], Suzuki et al. [37] and others, that a non-intrusive system, such as the NCSSS, or a video-based imaging recording, are likely better suited than a wearable or contact system for use in dementia patients in a falls detection study.

7.5 Discussion and Recommendations

7.5.1 Answer to Study Questions

This feasibility study is one of only a handful of Australian studies into the use of smart sensors for monitoring older adults in a residential care facility. The operational lessons learned in this real-world setting regarding feasibility and implementation of ambient sensors are significant.

Unexpected technical difficulties delayed full implementation of sensors in all participants' rooms. It was discovered that some rooms did not have live connections to the facility's network; this required the Technology partner to source an alternative method for networking the devices and collecting the sensor data. This reinforces the need to perform end to end live-testing of hardware and networks in each target room during the test phase, rather than relying on a test in a demonstration room of the facility.

As noted, the positions of the sensors required adjustment to overcome radio frequency interference in some rooms. Other studies have similarly reported on interference of smart sensor systems that required adaptation in their respective trials [34]. In the present pilot, positioning the sensors outside the optimal 4-m patient range caused deterioration in skeletal tracking performance. The accuracy of the pose-tracking element of the sensors is a key component in the system. The limitation was corrected by raising the detection level threshold resulting in the capture of a narrower range, so that the tracking was more optimised to a patient's movements within a smaller capture space of physical surroundings by the smart sensor. Clearly there is consideration in future studies to work with clinicians to reach a realistic calibration for the sensors which can balance sensitivity and specificity in falls and behaviour monitoring.

In addition, our investigation of staff and carer attitudes to the monitoring technology has shown they were overwhelmingly positive, especially for the residents in our cohort who had dementia or similar cognitive conditions. Carers and staff could envisage a range of benefits if the technology were proven to work. During post-trial interviews, staff nominated additional benefits such as knowing the events that lead to a fall, or gathering specific behaviour information that would enable the staff to be proactive. Staff felt the visualisations used in the depth camera images protected individual's privacy.

7.5.2 *Strengths and Limitations of the Study*

By its nature, the pilot study contained elements of value, as well as limitations. The conduct of the study in a real Australian residential care setting has provided an opportunity to test many facets of the NCSSS pilot, as well as engaging with participants' families or carers and clinical staff. It will now be possible to explore selected findings in more depth, for example by testing sensors for improved pose-tracking performance, by developing optimal network configurations for use where sensors are to be implemented, and by including more active or other participant group types.

The complexity of providing robust device connectivity, free from interference and within the specifications of sensor device capability (distance from patient in this case), plus, secure data storage and subsequent analysis of data recordings, cannot be underestimated in terms of resource and feasibility. A particular strength of the study is the reinforcement of this complexity in an assistive technology-based pilot.

The small sample size and short duration in this study are key weaknesses. Whilst it is accepted that recruitment of older participants for any study is more difficult than younger subjects, the present low numbers and brief duration of monitoring remain inconclusive about longer-term performance and efficacy of smart sensors for falls monitoring.

7.6 Future Research Directions

The ability of a NCSSS to accurately identify fall events or activities that may place people at risk of falls, such as walking to the bed unaccompanied, in a real life setting still needs to be established. The corresponding ability of a smart sensor system to reduce time between a fall and assistance being delivered to the patient also needs to be established through future studies. The existing literature has reached similar conclusions about the critical need for continuing research of falls detection and prevention assistive technologies and potential translation to implementation in care settings.

Existing monitoring at the care facility, namely pressure mat and the nurse call alert systems, was not disabled during this pilot study. However, the possibility that sensor system alerts could be integrated with alerts from existing systems is considered to be highly desirable. Vandenberg et al. [42] and Teh et al. [38] suggest that a coordinated communication system is essential for responding effectively to fall alerts.

Development of an integration approach with an intelligent system to combine, analyse and make sense of the underlying disparate systems data would be a valuable area for further research. Such an intelligent system could learn from the detected and subsequently human verified events, such that predictive ability is enhanced over time reducing false positive or negative alerts.

The current project further highlighted the importance of multidisciplinary research teams to investigate complex real world problems. While the project team had input from researchers with a range of skills, the investigators identified the need for a data technologist to advise during implementation on managing data pathways and data tracking, and to support data correlation across the existing data systems. As the sensors generated large quantities of data, specific skills are highly desirable to organise, transfer and store the raw material and suggest appropriate analytical methods. The role of team members in the delivery of an intelligent falls detection system also needs to take into account that multifactorial interventions could be effective in reducing fall rates in RACs, when additional external expertise and resources are provided in the short term [13].

As a final recommendation for future work, this study might prompt further exploration of new instruments in categorizing falls, such as the FARSEEING taxonomy that allows for comparative evaluation. The authors have considered elements of the study using the FARSEEING taxonomy for the purposes of describing the approach and technologies, as well as for supporting the analysis of results. The FARSEEING repository holds data for more than 300 falls, and the Consortium is actively seeking studies in this area [18].

7.7 Conclusion

The study was conducted to inform the feasibility and acceptability of a NCSSS to monitor behaviour and detect falls of older participants in a RAC setting. The findings have the potential to inform the development of a clinical framework to guide clinicians and healthcare providers in selection and use of non-invasive monitoring systems that may provide proactive alerting with the aim to prevent falls injury and/or falls events.

Critically, the study further points to the value of real patient case studies in healthcare research that can add to an evidence base of the patient experience. If validated, forms of non-contact sensor technology could enable speedier falls detection, particularly with the possibility of capturing the exact sequence of events preceding a fall, slip or trip that could support predictive analysis – both questions which offer benefit for individual and community falls prevention efforts.

Acknowledgements Study participants and their carers; Staff at the residential care facility; Professor Fernando Martin-Sanchez, Universidad de A Coruna; Advisory Board members Dr Frances Batchelor (National Ageing Research Institute) Melbourne, Prof George Demiris (Department of Biobehavioral Nursing and Health Systems, University of Washington, Seattle), Prof. Dr. Michael Marschollek (Hannover Medical School and University of Braunschweig—Institute of Technology); Dr Karen Courtney (School of Health Information Science, University of Victoria, Victoria, Canada) for permission to adapt her interview guide, and Raoul Ney, Semantrix Pty Ltd.
This study was funded in 2015 by the Networked Society Institute (formerly, Institute for a Broadband-enabled Society), The University of Melbourne, and completed in 2017.

References

1. Allan, L.M., Ballard, C.G., Rowan, E.N., Kenny, R.A.: Incidence and prediction of falls in dementia: a prospective study in older people. PLoS ONE **4**, e5521 (2009)
2. Aloulou, H., Mokhtari, M., Tiberghien, T., Biswas, J., Phua, C., Lin, J.H.K., Yap, P.: Deployment of assistive living technology in a nursing home environment: methods and lessons learned. BMC Med. Inform. Decis. Mak. **13**(42), 17 p (2013). https://doi.org/10.1186/1472-6947-13-42
3. Australian Institute of Health and Welfare: Admitted Patient Care 2015–16. Australian Hospital Statistics. Health Services Series No. 75. AIHW, Canberra (2017)
4. Boulton, E., Hawley-Hague, H., Vereijken, B., Clifford, A., Guldemond, N., Pfeiffer, K., Hall, A., Chesani, F., Mellone, S., Bourke, A., Todd, C.: Developing the FARSEEING Taxonomy of Technologies: Classification and Description of Technology Use (including ICT) in Falls Prevention Studies. J. Biomed. Inform. (2016). https://doi.org/10.1016/j.jbi.2016.03.017
5. Bradley, C.: Trends in Hospitalisations Due to Falls by Older People, Australia 1999-00 to 2010-11. Injury Research and Statistics No. 84. Cat. No. INJCAT 160. AIHW, Canberra (2013)
6. Brender, J., Talmon, J., de Keizer, N., Nykänen, P., Rigby, M., Ammenwerth, E.: STARE-HI—statement on reporting of evaluation studies in health informatics: explanation and elaboration. Appl. Clin. Inform. **4**(3), 331–358 (2013). https://doi.org/10.4338/aci-2013-04-ra-0024
7. Cameron, I.D., Gillespie, L.D., Robertson, M.C., Murray, G.R., Hill, K.D., Cumming, R.G., Kerse, N.: Interventions for preventing falls in older people in care facilities and hospitals. Cochrane Database Syst. Rev. **12**(Art. No.: CD005465) (2012). https://doi.org/10.1002/14651858.cd005465.pub3
8. Chaudhuri, S., Thompson, H., Demiris, G.: Fall detection devices and their use with older adults: a systematic review. J. Geriatr. Phys. Therapy **00**, 1–19 (2013). https://doi.org/10.1519/JPT.0b013e3182abe779
9. Farshchian, B.A., Dahl, Y.: The role of ICT in addressing the challenges of age-related falls: a research agenda based on a systematic mapping of the literature. Pers. Ubiquit. Comput. **19**(3–4), 649–666 (2015). https://doi.org/10.1007/s00779-015-0852-1
10. Feldwieser, F., Gietzelt, M., Goevercin, M., Marschollek, M., Meis, M., Winkelbach, S., Wolf, K.H., Spehr, J., Steinhagen-Thiessen, E.: Multimodal sensor-based fall detection within the domestic environment of elderly people. Zeitschrift für Gerontologie und Geriatrie **47**(8), 661–665 (2014). https://doi.org/10.1007/s00391-014-0805-8
11. Fischer, S.H., David, D., Crotty, B.H., Dierks, M., Safran, C.: Acceptance and use of health information technology by community-dwelling elders. Int. J. Med. Inform. **83**(9), 624–635 (2014). https://doi.org/10.1016/j.ijmedinf.2014.06.005
12. Fitzgerald, T.D., Hadjistavropoulos, T., Williams, J., et al.: The impact of fall risk assessment on nurse-led fears, patient falls and functional ability in long term care. Disabil. Rehabil. **38**(11), 1041–1052 (2016)
13. Francis-Coad, Jacqueline, Etherton-Beer, Christopher, Burton, Elissa, Naseri, Chiara, Hill, Anne-Marie: Effectiveness of complex falls prevention interventions in residential aged care settings: a systematic review. JBI Database Syst. Rev. Implementation Rep. **16**(4), 973–1002 (2018). https://doi.org/10.11124/JBISRIR-2017-003485
14. Gietzelt, M., Spehr, J., Ehmen, Y., Wegel, S., Feldwieser, F., Meis, M., Marschollek, M., Wolf, K.H., Steinhagen-Thiessen, E., Govercin, M.: GAL@Home: a feasibility study of sensor-based in-home fall detection. Zeitschrift für Gerontologie und Geriatrie **45**(8), 716–721 (2012). https://doi.org/10.1007/s00391-012-0400-9
15. Greenhalgh, T., Shaw, S., Wherton, J., Hughes, G., Lynch, J., A'Court, C., Hinder, S., Fahy, N., Byrne, E., Finlayson, A., Sorell, T., Procter, R., Stones, R.: SCALS: a fourth-generation study of assisted living technologies in their organisational, social, political and policy context. BMJ Open **6**(2), e010208 (2016). https://doi.org/10.1136/bmjopen-2015-010208
16. Hawley-Hague, H., Boulton, E., Hall, A., Pfeiffer, K., Todd, C.: Older adults' perceptions of technologies aimed at falls prevention, detection or monitoring: a systematic review. Int. J. Med. Informatics **83**(6), 416–426 (2014). https://doi.org/10.1016/j.ijmedinf.2014.03.002

17. Jancey, J., Wold, C., Meade, R., Sweeney, R., Davison, E., Leavy, J.: A balanced approach to falls prevention: application in the real world. Health Promot. J. Austral **00**, 1–5 (2018). https://doi.org/10.1002/hpja.42
18. Klenk, J., et al.: The FARSEEING real-world fall repository: a large-scale collaborative database to collect and share sensor signals from real-world falls. Eur. Rev. Aging Phys. Act. **13**, 8 (2016). https://doi.org/10.1186/s11556-016-0168-9
19. Kosse, N.M., Brands, K., Bauer, J.M., Hortobagyi, T., Lamoth, C.J.: Sensor technologies aiming at fall prevention in institutionalized old adults: a synthesis of current knowledge. Int. J. Med. Inform. **82**(9), 743–752 (2013). https://doi.org/10.1016/j.ijmedinf.2013.06.001
20. Lipsitz, L.A., Tchalla, A.E., Iloputaife, I., Gagnon, M., Dole, K., Su, Z.Z., Klickstein, L.: Evaluation of an automated falls detection device in nursing home residents. J. Am. Geriatr. Soc. **64**(2), 365–368 (2016). https://doi.org/10.1111/jgs.13708
21. Ludwig, W., Wolf, K.H., Duwenkamp, C., et al.: Health-enabling technologies for the elderly—an overview of services based on a literature review. Comput. Methods Programs Biomed. **106**(2), 70–78 (2012)
22. Marschollek, M., Becker, M., Bauer, J.M., Bente, P., Dasenbrock, L., Elbers, K., Hein, A., Kolb, G., Kunemund, H., Lammel-Polchau, C., Meis, M., Meyer Zu Schwabedissen, H., Remmers, H., Schulze, M., Steen, E.E., Thoben, W., Wang, J., Wolf, K.H., Haux, R.: Multimodal activity monitoring for home rehabilitation of geriatric fracture patients–feasibility and acceptance of sensor systems in the GAL-NATARS study. Inform. Health Soc. Care **39**(3–4), 262–271 (2014). https://doi.org/10.3109/17538157.2014.931852
23. Nijhof, N., van Gemert-Pijnen, L.J., Woolrych, R., Sixsmith, A.: An evaluation of preventive sensor technology for dementia care. J. Telemed. Telecare **19**(2), 95–100 (2013). https://doi.org/10.1258/jtt.2012.120605
24. Nunan, S., Wilson, C.B., Henwood, T., Parker, D.: Fall risk assessment tools for use among older adults in long-term care settings: a systematic review of the literature. Aust. J. Ageing **37**(1), 23–33 (2017). https://doi.org/10.1111/ajag.12476
25. Pang, I., Okubo, Y., Sturnieks, D., Lord, S.R., Brodie, M.A.: Detection of near falls using wearable devices. Syst. Rev. J. Geriatr. Phys. Therapy (2018). https://doi.org/10.1519/jpt.0000000000000181 (Epub ahead of print)
26. Peek, S.T., Wouters, E.J., van Hoof, J., Luijkx, K.G., Boeije, H.R., Vrijhoef, H.J.: Factors influencing acceptance of technology for aging in place: a systematic review. Int. J. Med. Inform. **83**(4), 235–248 (2014). https://doi.org/10.1016/j.ijmedinf.2014.01.004
27. Peetoom, K.K., Lexis, M.A., Joore, M., Dirksen, C.D., De Witte, L.P.: Literature review on monitoring technologies and their outcomes in independently living elderly people. Disabil. Rehabil. Assist Technol **10**(4), 271–294 (2015). https://doi.org/10.3109/17483107.2014.961179
28. Pol, M., Poerbodipoero, S., Robben, S., Daams, J., van Hartingsveldt, M., de Vos, R., de Rooij, S., Krose, B., Buurman, M.B.: Sensor monitoring to measure and support daily functioning for independently living older people: a systematic review and road map for further development. JAGS **61**(12), 219–227 (2013). https://doi.org/10.1111/jgs.12563
29. Potter, P., et al.: Evaluation of sensor technology to detect fall risk and prevent falls in acute care. Joint Commission J. Qual. Patient Safety **43**(8), 414–421 (2017)
30. Potter, P., Allen, K., Costantinou, E., Klinkenberg, D., Malen, J., Norris, T., O'Connor, E., Roney, W., Tymkew, H.H.: Anatomy of inpatient falls: examining fall events captured by depth-sensor technology. Joint Commission J. Qual. Patient Safety **42**(5), 225–231 (2016)
31. Rantz, M.J., Skubic, M., Miller, S.J., Galambos, C., Alexander, G., Keller, J., Popescu, M.: Sensor technology to support aging in place. J. Am. Med. Dir. Assoc. **14**(6), 386–391 (2013). https://doi.org/10.1016/j.jamda.2013.02.018
32. Rantz, M., Skubic, M., Abbott, C., Galambos, C., Popescu, M., Keller, J., Stone, E., Back, J., Miller, S.J., Petroski, G.F.: Automated in-home fall risk assessment and detection sensor system for elders. Gerontologist **55**(Suppl 1), S78–S87 (2015a). https://doi.org/10.1093/geront/gnv044

33. Rantz, M.J., Skubic, M., Popescu, M., Galambos, C., Koopman, R.J., Alexander, G.L., Phillips, L.J., Musterman, K., Back, J., Miller, S.J.: A new paradigm of technology-enabled 'vital signs' for early detection of health change for older adults. Gerontology **61**(3), 281–290 (2015b). https://doi.org/10.1159/000366518
34. Shinmoto Torres, R.L., Visvanathan, R., Abbott, D., Hill, K.D., Ranasinghe, D.C.: A battery-less and wireless wearable sensor system for identifying bed and chair exits in a pilot trial in hospitalized older people. PLoS ONE **12**(10), e0185670 (2017)
35. Stone, E.E., Skubic, M.: Fall detection in homes of older adults using the microsoft kinect. IEEE J. Biomed. Health Inform. **19**(1), 290–301 (2015). https://doi.org/10.1109/JBHI.2014.2312180
36. Stucki, R.A., Urwyler, P., Rampa, L., Muri, R., Mosimann, U.P., Nef, T.: A web-based non-intrusive ambient system to measure and classify activities of daily living. J. Med. Internet Res. **16**(7), e175 (2014). https://doi.org/10.2196/jmir.3465
37. Suzuki, R., Otake, S., Isutzu, T., Yoshida, M., Iwaya, T.: Monitoring daily living activities of elderly people in a nursing home using an infrared motion-detection system. Telemed. eHealth **12**(2), 146–156 (2006)
38. Teh, R.C., Mahajan, N., Visvanathan, R., Wilson, A.: Clinical effectiveness of and attitudes and beliefs of health professionals towards the use of health technology in falls prevention among older adults. Int. J. Evid. Based Healthcare **13**(4), 213–223 (2015). https://doi.org/10.1097/xeb.0000000000000029
39. Teh, R.C., Mahajan, N., Visvanathan, R., Ranasinghe, D., Wilson, A.: Evaluation and refinement of a handheld health information technology tool to support the timely update of bedside visual cues to prevent falls in hospitals. Int. J. Evidence Based Healthcare **15** (2017)
40. Tinetti, M., Kumar, C.: The patient who falls: "It's always a trade-off". JAMA **303**, 258–266 (2010)
41. Tovell, A., Harrison, J.E., Pointer, S.: Hospitalised Injury in Older Australians, 2011–12. Injury Research and Statistics Series No. 90. Cat. No. INJCAT 166. AIHW, Canberra (2014)
42. Vandenberg, A.E., van Beijnum, B.-J., Overdevest, V.G.P., Capezuti, E., Johnson Ii, T.M.: US and Dutch nurse experiences with fall prevention technology within nursing home environment and workflow: A qualitative study. Geriatr. Nurs. (2016). https://doi.org/10.1016/j.gerinurse.2016.11.005
43. Whitney, J., Close, J.C.T., Lord, S.R., Jackson, S.H.D.: Identification of high risk fallers among older people living in residential care facilities: a simple screen based on easily collectable measures. Arch. Gerontol. Geriatr. **55**, 690–695 (2012)
44. Wong Shee, A., Phillips, B., Hill, K., Dodd, K.: Feasibility, acceptability, and effectiveness of an electronic sensor bed/chair alarm in reducing falls in patients with cognitive impairment in a subacute ward. J. Nurs. Care Qual. **29**(3), 253–262 (2014). https://doi.org/10.1097/ncq.0000000000000054

Chapter 8
The Challenge of Automatic Eating Behaviour Analysis and Tracking

Dagmar M. Schuller and Björn W. Schuller

Abstract Computer-based tracking of eating behaviour is recently finding great interest by a broader choice of modalities such as by audio and video, or movement sensors, in particular in wearable every-day settings. Here, we provide an extensive insight into the current state-of-play for automatic tracking with a broader view on sensors and information used up to this point. The chapter is largely guided by and including results from the Interspeech 2015 Computational Paralinguistics Challenge (ComParE) Eating Sub-Challenge and the audio/visual Eating Analysis and Tracking (EAT) 2018 Challenge, both co-organised by the last author. The relevance is given by use-cases in health care and wellbeing including, amongst others, assistive technologies for individuals with eating disorders potentially leading either to under- or overeating or special health conditions such as diabetes. The chapter touches upon different feature representations including feature brute-forcing, bag-of-audio-word representations, and deep end-to-end learning from a raw sensor signal. It further reports on machine learning approaches used in the field including deep learning and conventional approaches. In the conclusion, the chapter discusses also usability aspects to foster optimal adherence, such as sensor placement, energy consumption, explainability, and privacy aspects.

Keywords Eating analysis · Eating disorders · mHealth · Health informatics · Deep learning · Artificial intelligence · Multimodality · Computational paralinguistics · Assistive technologies

D. M. Schuller · B. W. Schuller (✉)
audEERING GmbH, Gilching, Germany
e-mail: schuller@tum.de; schuller@ieee.org

B. W. Schuller
ZD.B Chair of Embedded Intelligence for Health Care and Wellbeing, University of Augsburg, Augsburg, Germany

GLAM - Group on Language, Audio & Music, Imperial College London, London, UK

H. Costin et al. (eds.), *Recent Advances in Intelligent Assistive Technologies: Paradigms and Applications*, Intelligent Systems Reference Library 170,
https://doi.org/10.1007/978-3-030-30817-9_8

8.1 Introduction

Eating is a major necessity that we usually undertake repeatedly every day. In fact, according to the popular US 2006 animated film "Over the Hedge" main character—a racoon—*"we eat to live"*. But, humans even *"eat when they've too much food!"*, and then, they exercise to get *"rid of the guilt so they can eat more food!"*, and ultimately, it all appears to be about *"food, food, food, food, FOOD!"*, as *"for humans, enough is never enough!"*. Likewise, it seems not surprising that the main concern of many or most of us has become food intake control. Unfortunately, wrong nutrition or eating patterns can lead to severe health and wellbeing problems. The World Health Organisation [34] estimates obesity levels have tripled across the globe since the authors' birth year of 1975. There is, however, also undereating disorders, such as Bulimia nervosa, which are equally life threatening. At the same time, diabetes prevalence has almost been increased by factor four since 1980 according to the World Health Organisation [35]. These trends also come at a significant psychological and economic burden both on individuals and societies. Triggered by these dramatic trends of malnutrition at large, efforts on research on automatically food intake monitoring are on the rise. In particular, the rise of assistive technologies has led to a plethora of approaches geared towards helping individuals to monitor their food intake and eating behaviour patterns such as in the recent challenges held in the field as described here and by Cummins et al. [4] based on the data collected by Hantke et al. [13]. Efforts are particularly made to include such assistive technology in wearables, as shown, for example, by Fontana et al. [8] to monitor ingestive behaviour [26].

The ultimate goal of such approaches could be to perfectly track primarily the amount and type of food taken in such as by (1) caloric intake, (2) type of food compatibility such as gluten-free, kosher, pescarian, paleo, vegan, vegetarian, etc., (3) macro-nutrient composition such as ratio of carbohydrates, fat, and proteins, and (4) micro-nutrient composition such as amount of saturated or non-saturated fats over certain time intervals such as on a day-to-day basis. As secondary parameters of interest for automated tracking, one could name timing and behaviour related aspects, such as (1) eating time patterns, and (2) speed, amount, force, and location of chewing and swallowing. Finally, qualitative and affective aspects may be of interest, such as (1) enjoyment, or (2) difficulty of food-intake.

With such information, one can greatly support mobile Health (mHealth) or other suited means of assistive technology to advise and coach—potentially in gamified ways—users such as when on diets limiting their daily caloric intake. Composition of food intake by type can, e.g., help in following specific diet patterns; macro-nutrients play a key role in diets such ketogenic or low-fat dietary plans; timing can assist in following fasting patterns such as in intermittent fasting rules, e.g., 16:8 h of noneating/eating per day or 5:2 days per week of eating/non-eating, etc.

Tracking the qualitative and quantitative aspects of chewing and swallowing may be used to aid in special dental care conditions or when facing larynx health issues, or simply to help improve one's eating behaviour such as slowing down one's eating and ensuring sufficient chewing, etc.

Finally, attitudinal and guiding information may help give advice about general eating behaviour, e.g., a wearable or phone app advising its user *"You seem to have only moderately enjoyed this quickly taken in pumpkin-soup? I estimate it to carry 64 kcals and 6.2 g carbohydrates, thus contributing to exceeding your daily limit. To ensure staying in ketosis – perhaps next time you may want to consider a Japanese miso soup which has 28 kcal and 2.6 g carbs for the same portion? It is also recommended that if you can wait for another hour, you can keep within your intermittent fasting regime."* (the nutritional values in the example were taken from http://fddb.info as accessed on 3 February 2019).

At the same time, a major goal of automatic eating behaviour monitoring obviously is to reach this kind of automated information assessment at the least possible effort for a user, which may be notoriously difficult if not close to unreachable as for some of these aspects. For example, caloric intake assessment will likely remain an estimate, albeit it can certainly be improved by uniting suited multiple sensors and intelligent solutions of machine learning and artificial intelligence.

Over the last decades, a plethora of approaches has thus been suggested in the literature, including mainly audio and video sensors, haptic and movement-related sensors, as well as more "exotic" solutions such as olfactory sensors, neuroimaging, and neuromodulation [32], among other techniques.

Here, our objectives are to (1) give a short general overview on solutions and current abilities in terms of automated eating behaviour tracking and analysis from a multisensorial and multimodal perspective in Sect. 8.2, to then (2) provide a deep insight into the state-of-play of audio (Sect. 8.3) and audio-visual (Sect. 8.4) eating behaviour monitoring based on the Interspeech 2015 Computational Paralinguistic Challenge's Eating Sub-Challenge held by Schuller et al. [28] and the EAT Challenge held at ACM ICMI in 2018 by Hantke et al. [14]—both of which were initiated and co-organised by the last author. Finally, (3) in Sect. 8.5, we highlight less targeted, yet highly relevant aspects such as adherence and usability, energy consumption, confidence measurement, or privacy enhancement as necessary before choosing a suitable enabling technology. We further provide some future research directions based on the above, followed by our conclusions.

8.2 Multisensorial and Multimodal Eating Behaviour Analysis in a Nutshell

Not only is one given a broad choice of sensors and modalities suited for the tracking of one's eating behaviour, but given their diverse shortcomings and strengths, and the rich list of desired parameters to ideally measure, it seems clear that the most can be gained by a reasonable combination of these. In this section, we will briefly introduce the diverse options for sensing and their characteristics supported by the relevant research literature.

8.2.1 Audio-Based Eating Analysis

Audio-based food intake monitoring and eating behaviour analysis bears great potential, as a microphone can easily be placed close to the target area such as in hearables (an electronic technology that is incorporated into a device or accessory worn in or on the ear usually consisting of a loud speaker, potentially added by sensors and some smart technology), or jewellery, etc. The sensors can be conductive such as bone conductors, passive, such as dynamic microphones, or active such as condenser microphones. For example, Alshurafa et al. [1] integrated a piezoelectric sensor into a necklace and are able to recognise various kinds of food from its capture. Further, Thomaz et al. [30] position the audio sensor on a wrist-mounted device, to analyse ambient sounds in-the-wild spotting eating activities. In a study with 20 subjects wearing the device on average 5 h during one "normal" day, they report an accuracy of 86.6% in a person-independent setting. In contrast, Gao et al. [9] suggest the usage of commodity Bluetooth headsets to detect eating episodes by analysing chewing sounds. They mention the limited sampling rate in such hardware as the major bottleneck. In their results, they report 94–95% accuracy in lab-conditions and 65–76% respectively for field-condition. They further state an improved accuracy of 77–94% by deep learning despite facing ambient noise. In principle, one can consider usage of multiple microphones or an array to improve the signal quality, which, however, is barely exploited in the audio-based eating tracking literature up to this point. We will deal in particular with audio as modality in Sect. 8.3.

8.2.2 Video-Based Eating Analysis

At first inspection, video appears as quite a powerful modality, as compared to audio only, to monitor aspects such as detailed food composition. However, a major disadvantage can be the requirement to have a sensor in such position that it can either see the food (such as on a plate), or the eating subject—ideally in everyday life situations without the need of user actions such as taking pictures of the food, etc., to increase the adherence and usage. This may change with the increasing spread of cameras in smart-homes or even public places potentially accessible via the Internet of Things. As an example of food image analysis, Kitamura et al. [17] attempted amongst others to analyse the food composition automatically from a still image taken by users. A further example of video analysis will be given in more detail in Sect. 8.4.

8.2.3 *Movement-Based Eating Analysis*

Similar to audio-sensing and as opposed to video-sensing, (non-video) movement-based analysis can often be realised in comfortable ways for the user such as by accelerometers, e.g., in bracelets or other jewellery. Likewise, a wrist-worn 3D-accerlormeter can help monitor "plate-to-mouth" movements [6]. Rahman et al. [25] further observe head movement measurement on (Google) glasses as promising approach to recognise eating. They argue that neck-worn accelerometer sensors risk being less comfortable. Bedri et al. [3] demonstrate alternatively the measurement by an in-ear jaw movement sensor. They argue that from the ear, among others, one can measure tongue activities and jaw motion and introduce an Outer Ear Interface based on infrared proximity sensors. These allow for estimation of the deformation within the ear as triggered by lower jaw movement. Their range of applications for this interface includes food intake monitoring.

8.2.4 *Further Sensors*

At the time, Brain-Computer Interfaces (BCI) seem largely unsuited for ubiquitous "in-the-wild" everyday eating behaviour analysis. However, they bear great potential, e.g., when it comes to the analysis of enjoyment, further affective connotation, or even related disorders, such as addiction, etc. [32]. A range of further sensors seem suited, such as the recently maturing olfactory and gustatory sensors, or such targeting the digestion behaviour, e.g., by so-called "invisibles", i.e., sensors being swallowed that transmit data from inside the body.

8.2.5 *Multimodal Eating Analysis*

For best results of tracking, the synergistic multimodal combination of sensors usually offers the best choice. If well implemented, it can further cover for missing data such as in the case of sensor failure or simply in case of visual occlusions, too noisy acoustic measurement, etc. A recent overview on multimodal fusion and techniques is given by Oviatt et al. [20]. As one exemplary case, Merck et al. [18] combine head and wrist motion as measured by (Google) glasses and a smartwatch worn on each arm with audio capture via a custom earbud microphone. Further, for annotation purposes, video was recorded. In a study with 6 participants recorded over 72 h, 89% recall are reported by using audio to identify meals, but motion sensing was needed to locate individual intakes.

8.3 Eating Condition in the Interspeech 2015 Computational Paralinguistic Challenge

Try this: Close your eyes next time you are eating. It is quite encouraging for an audio-based solution how rich the information audible to the human is when it comes to food intake. Not only can we hear the "crunchiness" of, say, a banana versus a cookie, the frequency and strength of chewing and swallowing, breath-taking, and further human sounds during food-intake, but also qualitative markers can be heard. So, the idea to exploit this information computationally has been attempted repeatedly as shown above. To benchmark these approaches, the annually held Interspeech Computational Paralinguistics Challenge (ComParE) featured in 2015 an Eating Condition Sub-Challenge.

8.3.1 Introduction of the Challenge

In this challenge, the type of food had to be determined from audio of speech under eating condition. It was based on the iHEARu-EAT database described in [13] that features speech under eating of six types of food (cf. Table 8.1). 30 subjects (15 f, 15 m; 26.1 ± 2.7 years) were recorded in a quiet, low reverberant office room. Speakers were recorded reading from the German version of the Aesop fable "The North Wind and the Sun." Each participant read the entire fable once completely per food type such as apple, banana, etc., as shown in Table 8.1.

Spontaneous narrative speech was centred on topics such as favourite travel destination, genre of music, or sports activity. The narratives were segmented manually matching roughly the length of the six pre-defined units in the read story. 1414 turns

Table 8.1 iHEARu-EAT Database as used in the Interspeech Computational Paralinguistics Challenge Eating Condition Sub-Challenge [28] and ICMI 2018 Eating Analysis and Tracking (EAT) Challenge [13]. 30 speakers are included (15 female and 15 male speakers; mean 26.1 ± 2.7 years). From each of these, read and spontaneous speech is included. The 1.4 k utterances total up to 2.9 h of audio/video recordings

# Instances	Train	Test	Sum
No food	140	70	210
Apple	140	56	196
Nectarine	133	63	196
Banana	140	70	210
Crisp	140	70	210
Biscuit	133	70	203
Haribo	119	70	189
Sum	945	469	1414

Fig. 8.1 Examples from the iHEARu-EAT database. Shown are two subjects (rows 1 and 2), and three different food types

and 2.9 h of speech (sampled at 16 kHz) are contained in the final dataset, where 1/7 of the speech files are spontaneous speech by design. As can be seen in Table 8.1, the data were split speaker-independently into a training set (20 speakers) and test set (10 speakers), stratified by age and gender. In the challenge, the food type (no food or one of the six food types) during speaking had to be recognised. Figure 8.1 shows an example of the video recorded alongside the audio.

The Interspeech 2015 Computational Paralinguistics Challenge had seen 57 registrations, 31 sites uploaded their results, and 25 participant papers were submitted, out of which 16 were accepted. Out of these, 7 dealt with the Eating Condition Sub-Challenge, as are summarised in Table 8.2. In this table, also a fusion by majority vote is shown on the three and six best participant engines. The table caption further lists the top results of teams which did not have an accepted paper in the proceedings and can hence not be referenced.

8.3.2 *Participation in the Challenge*

The baseline of the challenge was given by a feature brute-force approach: The openSMILE [7] toolkit's ComParE feature set that uses 65 Low-Level-Descriptors (LLDs), their delta coefficients and statistical functionals such as mean, maximum, etc., applied to these LLDs plus deltas leads to 6373 acoustic features. These were fed into a linear kernel Support Vector Machine (SVM) and trained with Sequential Minimal Optimisation. As optimal parameter was $C = 10^{-3}$ observed and applied.

The baseline together with the participants' results are shown in Table 8.2. The competition measure was Unweighted Average Recall (UAR), which is the sum of

Table 8.2 Summary of the Interspeech 2015 Computational Paralinguistics Challenge Eating Condition Sub-Challenge. Given are the contributions by author reference, the method, and the best Unweighted Average Recall (UAR) on the (subject independent) test set

Author Reference	Method	%UAR
Schuller et al. [28]	Baseline: SVM	65.9
Prasad and Gosh [24]	FS, HC, SVM	67.3
Wagner et al. [33]	FW, HC	67.6
Pir and Brown [22]	Group FS	67.9
Pellegrini [21]	SVM, NN	68.4
Kim et al. [16]	i-vector, ASR	74.6
Milde and Biemann [19]	RL by CNN, OOD	75.9
Kaya et al. [15]	FV w/CN	83.1
–	Fusion of 3 best systems	83.2
–	Fusion of 6 best systems	83.7

ASR Automatic Speech Recognition-based feature information, *CN* Cascaded Normalisation, *CNN* Convolutional Neural Network, *FS* Feature Selection, *FV* Fisher Vectors, *FW* Frequency Weighting, *HC* Hierarchical Classification, *NN* Neural Network, *OD* Out-of-domain Data, *RL* Representation Learning, *SVM* Support Vector Machines. Submissions without accepted papers are not listed in the table but the percentages of those teams who reached the top %UAR are as follows: 45.5, 60.7, 64.4, 65.1, 66.1, 68.9, and 73.6

the class-wise recall divided by the number of classes. This measure is well suited for imbalanced class-distributions. For seven classes, the chance level would be 1/7 $\approx$ 14.3% UAR.

The challenge winning contribution by Kaya et al. [15] reached 83.1% UAR. It was based on a novel combination of Fisher vector encoding of the extracted features and a cascaded normalisation to cope with variability due to different speakers and content.

The follow-up second best contribution by Milde and Biemann [19] employed representation learning by Convolutional Neural Networks (CNNs). Their CNN consisted of three convolutional and max-pooling layers, followed by two fully connected layers. During training, dropout, and data augmentation via pitch shifting were applied together with a transfer learning approach: The CNN was pre-trained on data from the Voxforge2 data. Finally, they trained logistic regression classifiers on the CNN features to reach the given 75.9% UAR.

These two participations represent the top end of the scores, and at the same time the two major current directions of modelling via 'traditional' approaches or deep learning—in particular, when it comes to representation of the audio. For the further entries into the challenge, Table 8.2 indicates the used methods focussing on the key characteristic aspect of each contribution. As can further be seen, the fusion of the n-best participant engines led only to a minor improvement, with the top score resembling 83.7% UAR. Further papers have since been based on the data from the challenge, such as by Gosztolya and Tóth [10].

8.4 EAT—The ICMI 2018 Eating Analysis and Tracking Challenge

Subsequently to the Interspeech 2015 ComParE Eating Sub-Challenge, the same database was featured in the EAT Challenge (Eating Analysis & Tracking—held at the ACM ICMI conference in 2018). In addition to the previous challenge, the video recording was given to the participants turning it into an audio-visual challenge. Furthermore, two new tasks were introduced, namely, likability of the food recognising the subjects' food likability rating as well as difficulty of eating as was judged by the participants in the study.

8.4.1 Introduction of the Challenge

As before, in a first Food-type Sub-Challenge, the six food-types versus no food had to be recognised automatically. In addition, the iHEARu-EAT database [13] contains beyond the food-type further target labels that have been collected as follows: After the recordings, participants self-reported on how much they liked each sort of food eaten by positioning a continuous slider between 0–*dislike extremely* and 1–*like extremely*. For the Likability Sub-Challenge, this was broken down into a binary task of *'Neutral'* or *'Like'*, where neutral was used in the case of no food, and dislike did not occur, as such food did not need to be eaten by the subjects. This binarisation took into account individual rating ranges. In addition, a 5-point Likert scale served to assess their difficulty of eating encountered per food type. In this case, the Difficulty Sub-Challenge was executed as a regression task.

Two different baseline systems were used in the challenge, as depicted in Figs. 8.2 and 8.3. The first baseline system as shown in Fig. 8.2 bases for audio on the same

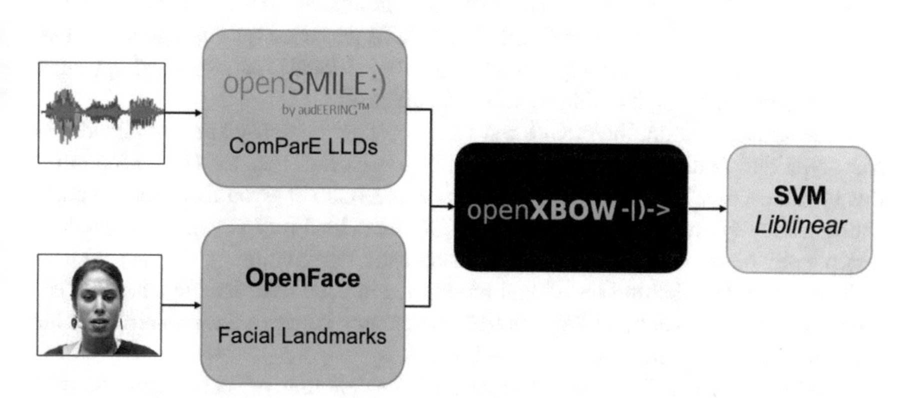

Fig. 8.2 Official baseline system of the iHEARu-EAT challenge. Audio features are extracted as Low-Level-Descriptors (LLDs)—65 in total—plus their deltas. Video features are facial landmarks. Cross-modal bag-of-words (XBOW) representations are then fed into a linear Support Vector Machine (SVM)

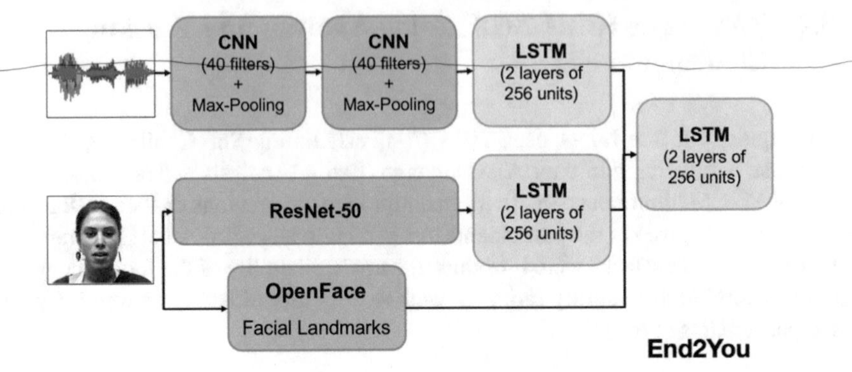

Fig. 8.3 Alternative baseline system of the iHEARu-EAT challenge. Convolutional Neural Networks (CNNs) are used both for audio and video representation learning. For video, a pre-trained network (ResNet-50) is used together with facial landmarks. Long-Short Term Memory (LSTM) recurrent neural networks are used for decision making

features as were already used in the Interspeech 2015 Challenge as described above, however, only using the LLDs of the ComParE feature set of openSMILE [7]. Subsequently, audio bag-of-words (BoWs) were extracted by openXBOW [27]. For video, OpenFace [2] provided 68 facial landmarks, which were then fed into openXBOW for video BoWs. The code-book size was optimised between 200 and 2000 entries—detailed results can be seen in [14]. The decision was again made by an SVM.

The second baseline system as shown in Fig. 8.3 uses end-to-end learning from raw audio and video. The tool used is end2you by Tzirakis et al. [31]. A mix of representation learning by CNNs, pre-trained CNNs, and the facial landmarks can be found as representation for Long-Short Term Memory (LSTM) recurrent neural networks.

Table 8.3 shows the results of these systems for audio-only, video-only, and audio-visual information exploitation on the test data. Alongside UAR for the classification tasks, Concordance Correlation Coefficient (CCC) was used for the regression task. One can easily observe the superiority of the 'traditional' feature-based approach over the representation learning by CNNs in all cases.

Interestingly though, the picture regarding optimal modality appears mixed: For food-type and (eating) difficulty, audio-only is superior, yet, for (food) likability, this is video-only. This does, however, appear reasonable, given that food-type and eating difficulty are largely acoustic tasks, whereas likability is valence-related—a task usually better solved with visual than acoustic information.

The fusion falls behind individual modalities in each case for the stronger 'traditional' baseline system—likely, as the discrepancy between the modalities is too high and additional information seems limited.

This is, however, different in the case of the 'alternative' deep representation learning approach, where in two out of three cases fusion leads to a gain.

Table 8.3 Given are (subject independent) test-set results per baseline system (openXBOW or End2You), Sub-Challenge, and modality (audio-only, video-only, or audio + video) on the iHEARu-EAT database as used in the 2018 EAT Challenge

Modality	openXBOW	End2You
Food-type Sub-Challenge: %UAR		
Audio-only	67.2	32.8
Video-only	27.0	24.5
Audio + Video	67.0	33.6
(Food) Likability Sub-Challenge: %UAR		
Audio-only	54.2	53.7
Video-only	58.3	50.9
Audio + Video	51.8	54.2
(Eating) Difficulty Sub-Challenge: CCC		
Audio-only	0.506	0.323
Video-only	0.252	0.220
Audio + Video	0.501	0.311

Measures: *UAR* Unweighted Average Recall, *CCC* Concordance Correlation Coefficient

8.4.2 Participation in the Challenge

Only four participating teams also passed the peer-review process and are quickly summarised as follows. Table 8.4 further provides a summary on the results. As can be seen, only in the Food-type Sub-Challenge, the baseline was surpassed by participants—hence, only this Sub-Challenge had an official winning contribution.

This winning contribution by Pir [23] introduces a Functional-based acoustic Group Feature Selection (FGFS) to select the functional-level features in groups per functional rather than individually. This optimisation of the feature space ultimately led to the best result of 69.1% UAR for food-type. Table 8.4 shows that the fusion of the four best systems led to 75.2% UAR for food-type, which however still falls clearly behind the best UAR of the Interspeech 2015 ComParE Eating Condition Sub-Challenge on the same data, but only with the audio available.

Sertolli et al. [29] were not official contestants, given that a co-organiser participated in this team. They applied pre-trained CNNs using the Wav2Letter net pre-trained on speech for end-to-end Automatic Speech Recognition (ASR) as feature extractor. Compact Bilinear Pooling (CBP) was further introduced to combine multiple feature representations extracted from different layers of the CNN. For modelling, a Recurrent Neural Network (RNN) classifier was used reaching 73.3% UAR for food-type—the highest result in this challenge.

Guo et al.'s [11] entry used a 4-level Double-Tree Complex Wavelet Transform Decomposition (DTCWTD) of an audio signal into five sub-audio signals from low to high frequency. From these, in addition to the 'traditional' functional-based features also representations of pretrained CNNs based on SliCQ-nonstationary Gabor

Table 8.4 Summary of the 2018 EAT Challenge. Given are the contributions by author reference, the method, and the best Unweighted Average Accuracy (UAR) and Concordance Correlation Coefficient (CCC) on the test set

Author Reference	Method	%UAR/CCC
Food-type Sub-Challenge: %UAR		
Hantke et al. [14]	Baseline: BoW	67.2
Haider et al. [12]	AFT + AFS	32.7
Guo et al. [11]	DTCWTD, FCTs + CNN + BoW	65.9
Pir [23]	FGFS	69.1
Sertolli et al. [29]	CNN + CBP, RNN	73.3
–	Fusion of 3 best systems	74.4
–	Fusion of 4 best systems	75.2
–	Fusion of 5 best systems	72.9
(Food) Likability Sub-Challenge: %UAR		
Hantke et al. [14]	Baseline: BoW	58.3
Pir [23]	FGFS	53.2
Haider et al. [12]	AFT + AFS	54.6
Guo et al. [11]	DTCWTD, FCTs + CNN + BoW	56.4
–	Fusion of 3 best systems	58.2
–	Fusion of 4 best systems	58.4
(Eating) Difficulty Sub-Challenge: CCC		
Hantke et al. [14]	Baseline: BoW	0.506
Guo et al. [11]	DTCWTD, FCTs + CNN + BoW	0.447
Haider et al. [12]	AFT + AFS	0.462
–	Fusion of 3 best systems	0.515

AFS Active Feature Selection, *AFT* Active Feature Transformation, *BoW* Bags-of-Words, *CBP* Compact Bilinear Pooling, *CNN* Convolutional Neural Network, *DTCWTD* Double-Tree Complex Wavelet Transform Decomposition, *FCTs* Functionals, *FGFS* Functional-based acoustic Group Feature Selection, *RNN* Recurrent Neural Network

transform and a cochleagram map, were computed in addition to the baseline Bag-of-Audio-Words features. Finally, the early fusion of all these three kinds of features yielded the best results.

Finally, Haider et al. [12] suggest Active Feature Transformation (AFT) of the ComParE features and Active Feature Selection (AFS). They further compare to Principal Component Analysis (PCA) for feature transformation. Ultimately, their reduced feature sets prevail, and AFS outperforms PCA and AFT for audio, and AFT of the (visual) facial landmarks) falls behind, whereas a weighted score fusion leads to better results.

8.5 Discussion and Recommendations: From Adherence to Opportunities

The above sections have shown that automatic eating behaviour tracking appears largely possible, if one can accommodate a certain margin of error. This seems, however, acceptable just like wearable activity trackers these days underlie a certain error but are largely used by the masses. In particular the challenges described above have shown that audio can be a quite reliable modality, but movement sensors can support it to track certain eating movements.

Video on the other hand holds the promise of food composition estimation, which, however, still seems a challenging task. In the last part of this chapter, we want to particularly touch upon a number of further aspects to ensure broad usage of relevant technology in the near future.

8.5.1 Adherence

Without adherence, all efforts would be in vain, as users would simply not stick to using the technology. Hence, a major aspect of adherence will be the usability of appropriate sensors and the software accessible to users. This will, however, also impact on the selection of the sensor and how easily it can be placed in addition to appearance and comfort questions. Further considerations will include whether such sensors might already be available in consumer devices used broadly such that no additional sensors will be needed. At present, this seems to favour monitoring by audio or movement from accelerometers, but of course, video can be an option as well.

8.5.2 Energy and Efficiency

The lion's share of research papers dealing with eating tracking analysis is focussing on recognition rate improvement. However, if continuous monitoring is to be used in a real-world ubiquitous mobile setting, energy consumption as well as processor and memory load become major points of concern, if, for example, run on smart devices that also fulfil other purposes for the user. This can be eased by cascaded processing where in a first stage low computational power is used for tracking eating activity onset accepting also lower accuracies. Once an onset candidate is spotted, it can be switched to higher accuracy, and accordingly higher load requirements. An interesting alternative can be the exploitation of multimodal information such as movement tracking by accelerometer for typical "plate to mouth" movements, before starting higher energy consumption and processing requirement modalities. Additionally, methods of energy and memory-efficient machine learning such as low-precision neural networks that employ quantified parameters, e.g., binary or ternary NNs, can of course be considered.

8.5.3 *Confidence Measurement and Explainability*

A further so far broadly ignored but crucial aspect is the computation of meaningful confidence measures alongside the eating behaviour tracking. This seems crucial, as the recognition itself is not fully reliable, as, e.g., shown by the results of the two competitive challenge events reported upon above. Such confidence measurement is best executed independent of the estimation of interest itself and can also be learnt in semi-supervised ways [5].

Beyond, explainability on a broader scale is an increasing concern of machine learning and pattern recognition systems these days. With the advent of deep learning as shown in the challenge contributions above, more interest can be expected in explainability—in particular again in the case where recognition and tracking results seem to be perceivably "off" to users.

8.5.4 *Privacy*

Privacy concerns if data is not processed on a private device but distributed to a server include in particular the potential presence of other individuals when audio or video is being recorded and transmitted. However, also eating behaviour data itself will need to be protected. The BoW approach can be a suited mean of feature information compression—the above presented results show that it can lead to state-of-the-art results despite highly compressed information representation. However, further means of privacy protection will need to be invested into.

8.5.5 *Opportunities on the Technical End*

Beyond the above highlighted aspects remains a number of further research directions which may lead to improved accuracies and user experiences. A major issue in machine learning in general is the often high desire in learning data—this holds in particular for deep learning approaches where up to millions of free learning parameters need to be learnt from data. While one may collect large databases of users at scale with suited services, it will always remain challenging to adapt to an individual user from little data. In the challenges reported above, it has also repeatedly been evidenced that usage of pre-trained networks can ease the lack of domain data. However, in addition, coupling analysis with synthesis of data can be promising. This is recently met, e.g., in the (generative) adversarial network (GAN) topologies that allow a neural network to synthesise its own training material based on initial examples.

Furthermore, rather than explicitly asking for user labelling as outlined above, e.g., by active learning, reinforcement learning holds the promise to allow for efficient weakly supervised learning.

Finally, automatic machine learning such as self-learning of suited network topologies may help to completely self-learn—potentially even end-to-end from ever new sensors' raw data—without human involvement.

8.5.6 Conclusion

In summary, the automatic tracking of eating behaviour used in everyday life based on a richer selection of sensors and being sufficiently reliable to be meaningful seems around the corner. Yet, clinical or health trials with patients using any of these technologies will be needed to translate into real world use in healthcare to prevent and treat eating disorders. To best answer how far we are to get there and what might still be needed to be done, practitioners, care-takers, and patients or users need to be included in the discussion and further studies and competitions need to be prepared and carried out. For example, the results of the two challenge events described here were not designed to come to conclusions about particular eating disorders and/or patient profiles/characteristics which may be more readily addressed by these assistive technologies in the future.

Then, with the described further improvements in accuracy, but also usability, efficiency, security, and reliability as well as explainability, one can expect such automated tracking to become a valuable assistive asset in improving our daily eating behaviours—and health and wellbeing into the future.

References

1. Alshurafa, N., Kalantarian, H., Pourhomayoun, M., Liu, J.J., Sarin, S., Shahbazi, B., Sarrafzadeh, M.: Recognition of nutrition intake using time-frequency decomposition in a wearable necklace using a piezoelectric sensor. IEEE Sens. J. **15**(7), 3909–3916 (2015)
2. Baltrušaitis, T., Robinson, P., Morency, L.P.: OpenFace: An open source facial behavior analysis toolkit. In: Proceedings IEEE Winter Conference on Applications of Computer Vision (WACV), Lake Placid, NY, USA, pp. 1–10 (2016)
3. Bedri, A., Byrd, D., Presti, P., Sahni, H., Gue, Z., Starner, T.: Stick it in your ear: building an in-ear jaw movement sensor. In: Adjunct Proceedings of the 2015 ACM International Joint Conference on Pervasive and Ubiquitous Computing and Proceedings of the 2015 ACM International Symposium on Wearable Computers, Osaka, Japan, pp. 1333–1338 (2015)
4. Cummins, N., Schuller, B.W., Baird, A.: Speech analysis for health: current state-of-the-art and the increasing impact of deep learning. Methods **151**, 41–54 (2018). (Special Issue on Translational data analytics and health informatics)

5. Deng, J., Schuller, B.: Confidence measures in speech emotion recognition based on semi-supervised learning. In: Proceedings Thirteenth Annual Conference of the International Speech Communication Association (Interspeech), Portland, OR, USA, pp. 2226–2229 (2012)
6. Drennan, M.: An assessment of linear wrist motion during the taking of a bite of food. Ph.D. Thesis. Clemson University, Clemson, SC, USA (2010)
7. Eyben, F., Weninger, F., Gross, F., Schuller, B.: Recent developments in openSMILE, the Munich open-source multimedia feature extractor. In: Proceedings 21st ACM International Conference on Multimedia, Barcelona, Spain, pp. 835–838 (2013)
8. Fontana, J.M., Farooq, M., Sazonov, E.: Automatic ingestion monitor: a novel wearable device for monitoring of ingestive behaviour. IEEE Trans. Biomed. Eng. **61**(6), 1772–1779 (2014)
9. Gao, Y., Zhang, N., Wang, H., Ding, X., Ye, X., Chen, G., Cao, Y.: iHear food: eating detection using commodity bluetooth headsets. In: Proceedings IEEE First International Conference on Connected Health: Applications, Systems and Engineering Technologies (CHASE), Washington, DC, USA, pp. 163–172 (2016)
10. Gosztolya, G., Tóth, L.: A feature selection-based speaker clustering method for paralinguistic tasks. Pattern Anal. Appl. **21**(1), 193–204 (2018)
11. Guo, Y., Han, J., Zhang, Z., Schuller, B., Ma, Y.: Exploring a new method for food likability rating based on DT-CWT theory. In: Proceedings 20th ACM International Conference on Multimodal Interaction (ICMI), Boulder, Colorado, pp. 569–573 (2018)
12. Haider, F., Pollak, S., Zarogianni, E., Luz, S.: SAAMEAT: active feature transformation and selection methods for the recognition of user eating conditions. In: Proceedings 20th ACM International Conference on Multimodal Interaction (ICMI), Boulder, Colorado, pp. 564–568 (2018)
13. Hantke, S., Weninger, F., Kurle, R., Ringeval, F., Batliner, A., El-Desoky Mousa, A., Schuller, B.: I hear you eat and speak: automatic recognition of eating condition and food types, use-cases, and impact on ASR performance. PLoS ONE **11**(5), 1–24 (2016)
14. Hantke, S., Schmitt, M., Tzirakis, P., Schuller, B.: EAT—the ICMI 2018 eating analysis and tracking challenge. In: Proceedings 20th ACM International Conference on Multimodal Interaction (ICMI), Boulder, Colorado, pp. 569–563 (2018)
15. Kaya, H., Karpov, A.A., Salah, A.A.: Fisher vectors with cascaded normalization for paralinguistic analysis. In: Proceedings Sixteenth Annual Conference of the International Speech Communication Association (Interspeech), Dresden, Germany, pp. 909–913 (2015)
16. Kim, J., Nasir, M., Gupta, R., van Segbroeck, M., Bone, D., Black, M.P., Skordilis, Z.I., Yang, Z., Georgiou, P.G., Narayanan, S.S.: Automatic estimation of Parkinson's disease severity from diverse speech tasks. In: Proceedings Sixteenth Annual Conference of the International Speech Communication Association (Interspeech), Dresden, Germany, pp. 914–918 (2015)
17. Kitamura, K., de Silva, C., Yamasaki, T., Aizawa, K.: Image processing based approach to food balance analysis for personal food logging. In: Proceedings IEEE International Conference on Multimedia and Expo (ICME), Singapore, pp. 625–630 (2010)
18. Merck, C., Maher, C., Mirtchouk, M., Zheng, M., Huang, Y., Kleinberg, S.: Multimodality sensing for eating recognition. In: Proceedings 10th EAI International Conference on Pervasive Computing Technologies for Healthcare, Cancun, Mexico, pp. 130–137 (2016)
19. Milde, B., Biemann, C.: Using representation learning and out-of-domain data for a paralinguistic speech task. In: Proceedings Sixteenth Annual Conference of the International Speech Communication Association (Interspeech), Dresden, Germany, pp. 904–908 (2015)
20. Oviatt, S., Schuller, B., Cohen, P., Sonntag, D., Potamianos, G.: The Handbook of Multimodal-Multisensor Interfaces: Signal Processing, Architectures, and Detection of Emotion and Cognition, vol. 2. Morgan & Claypool (2018)
21. Pellegrini, T.: Comparing SVM, Softmax, and shallow neural networks for eating condition classification. In: Proceedings Sixteenth Annual Conference of the International Speech Communication Association (Interspeech), Dresden, Germany, pp. 899–903 (2015)
22. Pir, D., Brown, T.: Acoustic group feature selection using wrapper method for automatic eating condition recognition. In: Proceedings Sixteenth Annual Conference of the International Speech Communication Association (Interspeech), Dresden, Germany, pp. 894–898 (2015)

23. Pir, D.: Functional-based acoustic group feature selection for automatic recognition of eating condition. In: Proceedings 20th ACM International Conference on Multimodal Interaction (ICMI), Boulder, Colorado, pp. 579–583 (2018)
24. Prasad, A., Gosh, P.K.: Automatic classification of eating conditions from speech using acoustic feature selection and a set of hierarchical support vector machine classifiers. In: Proceedings Sixteenth Annual Conference of the International Speech Communication Association (Interspeech), Dresden, Germany, pp. 884–888 (2015)
25. Rahman, S.A., Merck, C., Huang, Y., Kleinberg, S.: Unintrusive eating recognition using Google Glass. In: Proceedings IEEE 9th International Conference Pervasive Computing Technologies for Healthcare (PervasiveHealth), Istanbul, Turkey, pp. 108–111 (2015)
26. Sazonov, E.S., Makeyev, O., Schuckers, S., Lopez-Meyer, P., Melanson, E.L., Neuman, M.R.: Automatic detection of swallowing events by acoustical means for applications of monitoring of ingestive behaviour. IEEE Trans. Biomed. Eng. **57**(3), 626–633 (2010)
27. Schmitt, M., Schuller, B.: OpenXBOW: introducing the Passau open-source crossmodal bag-of-words toolkit. J. Mach. Learn. Res. **18**(1), 3370–3374 (2017)
28. Schuller, B., Steidl, S., Batliner, A., Hantke, S., Hönig, F., Orozco-Arroyave, J. R., Nöth, E., Zhang, Y, Weninger, F.: The INTERSPEECH 2015 computational paralinguistics challenge: nativeness, Parkinson's & eating condition. In: Proceedings Sixteenth Annual Conference of the International Speech Communication Association (Interspeech), Dresden, Germany, pp. 478–482 (2015)
29. Sertolli, B., Cummins, N., Sengur, A., Schuller, B.: Deep end-to-end representation learning for food type recognition from speech. In: Proceedings 20th ACM International Conference on Multimodal Interaction (ICMI), Boulder, Colorado, pp. 574–578 (2018)
30. Thomaz, E., Zhang, C., Essa, I., Abowd, G.D.: Inferring meal eating activities in real world settings from ambient sounds: a feasibility study. In: Proceedings 20th ACM International Conference on Intelligent User Interfaces (IUI), Atlanta, GA, USA, pp. 427–431 (2015)
31. Tzirakis, P., Zafeiriou, S., Schuller, B.W.: End2You—The Imperial Toolkit for Multimodal Profiling by End-to-End Learning (2018). arXiv preprint arXiv:1802.01115
32. Val-Laillet, D., Aarts, E., Weber, B., Ferrari, M., Quaresima, V., Stoeckel, L.E., Alonso-Alonso, M., Audette, M., Malbert, C.H., Stice, E.: Neuroimaging and neuromodulation approaches to study eating behavior and prevent and treat eating disorders and obesity. Neuro Image Clin. **8**, 1–31 (2015)
33. Wagner, J., Seiderer, A., Lingenfelser, F., André, E.: Combining hierarchical classification with frequency weighting for the recognition of eating conditions. In: Proceedings Sixteenth Annual Conference of the International Speech Communication Association (Interspeech), Dresden, Germany, pp. 889–893 (2015)
34. World Health Organization: Obesity and Overweight (2018). https://www.who.int/news-room/fact-sheets/detail/obesity-and-overweight
35. World Health Organization: Diabetes (2019). https://www.who.int/news-room/fact-sheets/detail/diabetes. Accessed 3 Feb 2019

Additional Reading Section (Resource List)

36. Amft, O., Junker, H., Troster, G.: Detection of eating and drinking arm gestures using inertial body-worn sensors. In: Proceedings Ninth IEEE International Symposium on Wearable Computers (ISWC), Osaka, Japan, pp. 160–163 (2005)
37. Dong, Y., Scisco, J., Wilson, M., Muth, E., Hoover, A.: Detecting periods of eating during free-living by tracking wrist motion. IEEE J. Biomed. Health Inf. **18**(4), 1253–1260 (2014)
38. Liu, J., Johns, E., Atallah, L., Pettitt, C., Lo, B., Frost, G., Yang, G.Z.: An intelligent food-intake monitoring system using wearable sensors. In: 2012 Ninth IEEE International Conference on Wearable and Implantable Body Sensor Networks (BSN), London, UK, pp. 154–160 (2012)

39. Mirtchouk, M., Merck, C., Kleinberg, S.: Automated estimation of food type and amount consumed from body-worn audio and motion sensors. In: Proceedings 2016 ACM International Joint Conference on Pervasive and Ubiquitous Computing (UbiComp), Heidelberg, Germany, pp. 451–462 (2016)
40. Nguyen, D.T., Cohen, E., Pourhomayoun, M., Alshurafa, N.: SwallowNet: recurrent neural network detects and characterizes eating patterns. In: Proceedings IEEE International Conference on Pervasive Computing and Communications Workshops, PerCom Workshops, Kona, HI, USA, pp. 401–406 (2017)
41. Schuller, B., Batliner, A.: Computational Paralinguistics. John Wiley & Sons (2013)

Printed in the United States
By Bookmasters